CONTENTS

PREFACE

The animal production industry relies heavily on a knowledge and understanding of animal physiology. This understanding allows producers to manipulate animals and so produce food efficiently. Animal production is concerned with the management of species which have been selected for domestication because they provide us primarily with a food source (meat, fish, eggs and milk). This book is about how we have used our knowledge of science to manipulate these animal species and improve productivity.

We start by looking at the important relationship we have with domestic animals – as a source of food and many other products such as textiles. We examine recent trends in meat consumption and compare the energetics of animal production with those of producing crops. Systems of animal production are discussed and these are compared with respect to animal welfare, productivity and economics.

The methods used by producers to improve domestic animal species through breeding programmes are also reviewed. This is expanded to include modern techniques such as embryo and gene transfer and artificial insemination.

We then study the production of three vertebrate species in detail. We begin by looking at cattle production and discuss how the reproductive cycle of the cow can be managed by the producer to provide beef and milk for the consumer. We examine the poultry industry and discuss both meat and egg production. The composition and quality of the egg is discussed and systems of production are compared. We also discuss the fish industry. We look at sea-fishing methods and the problems caused by overfishing. We examine techniques available to reduce the risks of overfishing, both through regulations and the development of fish-farms (aquaculture).

Finally, we discuss the dual role of the invertebrate as both producer and pest. We examine in detail the honey-bee as an example of a food producer and the aphid as an example of a crop pest.

This book aims to cover the material found in many of the option modules in Biology at A level but the material is also suitable for use by students studying vocational courses at similar levels. The material is also suitable for inclusion in General Studies courses. It is expected that students already have a knowledge of basic animal physiology and anatomy before studying the option modules covered by this book. In particular a knowledge of the following would be an advantage:

- general mammalian reproductive structure,
- formation of male and female gametes by meiosis,
- sexual reproduction in mammals,
- basic Mendelian genetics.

FOCUS ON BIOLOGY

ANIMAL SCIENCE IN ACTION

Caroline Barnes

for the
University of Nottingham

Hodder & Stoughton

A MEMBER OF THE HODDER HEADLINE GROUP

ACKNOWLEDGEMENTS

The original idea to increase awareness of Agriculture and Horticulture as applied biological sciences in schools and colleges was conceived through discussions between the National Farmers' Union and Department of Agriculture and Horticulture, at the University of Nottingham, both of whom initially financed the venture. Invaluable support was also given by the LSA Charitable Trust who funded Caroline Barnes to complete this book. All of these organisations have our grateful thanks.

Special thanks must also go to Dr Jeff Atherton of the University of Nottingham who had the original ideas for this book and *Plant Science in Action*, another book in this series. Without the funding and assistance that he secured, production of the books would not have been possible.

Many members of staff at the University freely gave data, observations and numerous examples for inclusion in this text. Particular thanks is due to Rob Clarke, Will Haresign, Phil Garnsworthy, Keith Scott and Julian Wiseman. Special thanks is also due to Mr D. Morris of Stonea Farm, Cambs., who gave invaluable help with information about fish-farming, and Mr C. Barnes for his help with the agricultural content of this book.

The publishers and the author are grateful to the following companies, institutions and individuals who have given permission to reproduce photographs in this book:
AFRC, Institute of Animal Physiology and Genetics Research, Cambridge (24 right); Alfred Pasieka/ Science Photo Library (49 top); Biophoto Associates (49 bottom): Bruce Coleman (5, 6 right, 44 top, 95 bottom right); Don Fawcett/Science Photo Library (39); Genus (16); Holt Studios International (10 right); John Howard/Science Photo Library (54); Oxford Scientific Films (59 number 2, 76 middle right, 79, 93 all, 94 top right, 95 left); Poultry World (76 bottom right); Roddy Paine (2 right, 6 left, 10 left, 13 all, 22 all, 24 left, 38, 45, 51 all, 59 all except number 2, 72, 75, 76 left and top right, 83, 88, 89 all left, 94 left, 95 top right, 96 both); Roslin Institute, Edinburgh (67 both, 69); Science Photo Library (100); Simon Fraser/Science Photo Library (89 right); Technical Services and Supplies (70); Zefa (1, 2 left, 9, 44 both middles and bottom, 94 bottom right, 97).

All artwork was drawn by Chartwell Illustrators.

We would also like to thank the following bodies who gave permission to reproduce copyright question material:
Northern Examinations Assessment Board, University of London Examinations and Assessment Council, and the Associated Examining Board.

For Christopher, Matthew and Oliver.

British Library Cataloguing in Publication Data

ISBN 0 340 60101 9

First published 1995
Impression number   10 9 8 7 6 5 4 3 2 1
                       1999 1998 1997 1996 1995

Typeset by Litho Link Ltd, Welshpool, Powys, Wales.
Printed in Great Britain for Hodder & Stoughton Educational, a division of Hodder Headline Plc, 338 Euston Road, London NW1 3BH by Bath Press Ltd.

1 HUMANS AND ANIMALS

LEARNING OUTCOMES

After studying this chapter you should be able to:
- describe the ways in which humans use other animal species, and be able to discuss the reasons why they do and know the arguments which say they should not,
- compare trends in the consumption of meat and animal products throughout the world and discuss the reasons for the changes which have been observed in recent years,
- describe the stages leading to the domestication of an animal species and list the characteristics which an animal suited to domestication may possess,
- talk about the problem of energy transfer in animal production which arises because so much energy is lost during normal metabolic processes,
- outline the three main systems used to produce animals for human consumption and evaluate the merits of each.

1.1 HUMANS AND ANIMALS

Since the Stone Age we have used animals to provide food, materials for clothing, drugs, fertilisers and transport. The majority of animal products which are important to us come from **domestic** species. This book is primarily concerned with the production of animals and animal products for food. Table 1.1 shows the use we as humans make of some animal species.

Animal products are commonly used for food by most cultures

Table 1.1   The use of domesticated animal species

SPECIES	MILK	FIBRE	MEAT	FERTILISER	HIDE	WOOL	TRACTION	TRANSPORT
Cattle	*	*	*	*	*		*	*
Sheep	*	*	*	*	*	*		
Horse	*	*	*	*	*		*	*
Goat	*	*	*	*	*			
Mule			*				*	*
Yak	*	*	*		*		*	*
Goose		*	*					
Rabbit			*		*			
Reindeer	*		*		*		*	*
Kangaroo			*		*			

* indicates a common use.   (Adapted from H N Turner, 1972 *Outlook on Agriculture 6*)

Many materials used for clothing come from animals

1.2 ANIMALS FOR FOOD

1.2.1 Meat consumption

Human beings have always hunted and killed animals for meat. Even today a range of species are bred and killed for this purpose. Table 1.2 shows the amount of meat and fish produced each year worldwide. It shows figures for the most commonly used species, although in certain regions animals such are reindeer, guinea pig and rabbit are also common sources of meat.

Table 1.2   Common sources of meat

SPECIES	MEAT PRODUCTION PER YEAR (10^4 tonnes)
Cattle	39 953
Sheep and goats	7006
Pigs	40 644
Horse	416
Poultry	19 818
Fish and other aquatic animals	65 700

(From FAO Data)

1.2.2 Animal products

Animal products, like meat, milk, eggs and the foods produced from them, make up a large part of the human diet. Cows are the main source of milk for human consumption throughout the world, although milk from other species, especially sheep and goats is also popular in many countries. Milk is an excellent food source because it is rich in protein, energy and minerals. Table 1.3 shows the percentage of the recommended daily intake (RDI) of several nutrients that are supplied by half a litre of cow's milk. The figures are based on the RDI for a growing child of five.

Table 1.3   Percentage of recommended daily intake for a five year old supplied by 0.5 l of cow's milk

NUTRIENT	% OF RDI
Calcium	70
Energy	25
Protein	40
Vitamin A	33
Thiamin (vitamin B_2)	33
Riboflavin (vitamin B_2)	70

However, milk has its problems. It is made up of about 88% water and so is bulky to transport from the farm to the consumer and to store. It also goes off very easily because it provides an excellent food source for bacteria as well as humans! Processing and converting milk into products such as cheese and butter helps to prolong the shelf life a little, as we shall discuss in a later chapter.

Milk products such as cheese can help prolong the shelf life of milk

Eggs also provide an excellent source of nutrients, and they have a longer shelf life than milk. Most eggs consumed by humans come from chickens although we also eat duck and quail's eggs. We shall talk further about egg production in a later chapter.

1.2.3 The consumption of meat worldwide

During the twentieth century there has been a steady increase in the consumption of animal products throughout North America and the countries in the west of Europe. The increase in consumption is due partly to an increase in the average income in these countries which has allowed people to buy more expensive foods, like meat and related animal products.

Until the 1990s, the countries in Eastern Europe had centrally controlled economies which made animal products artificially expensive and out of the reach of the majority of people. The diet of most people living in these regions was, therefore, based mainly on vegetables and cereals. However, since the start of the 1990s, shifts toward a market economy and the reorganisation of political systems within these regions has meant meat has become more available, and led to an increase in demand for animal products.

The developing regions of the world, like Africa, Asia and Latin America tend to consume low levels of meat and animal products. This is because these foods are relatively expensive and unaffordable on the low incomes which many families in these regions receive. Many people grow their own food and use it as 'barter'. Such populations rely mainly on plants for food. Their diet is composed of cereals, pulses and root crops — relatively cheap sources of energy, protein and other nutrients. Figure 1 shows the contribution made by animal proteins and plant proteins to the human diet in different regions of the world. You can see how the proportion made by animal and plant products varies from developed to developing regions.

1.2.4 Changing trends in meat consumption

Since the 1980s there has been a change in the trends of meat consumption. There has been a slight decrease in the amount of red meat, eggs and butter eaten in the more developed regions of the world. Recent health scares about food safety and diet have largely contributed to this trend.

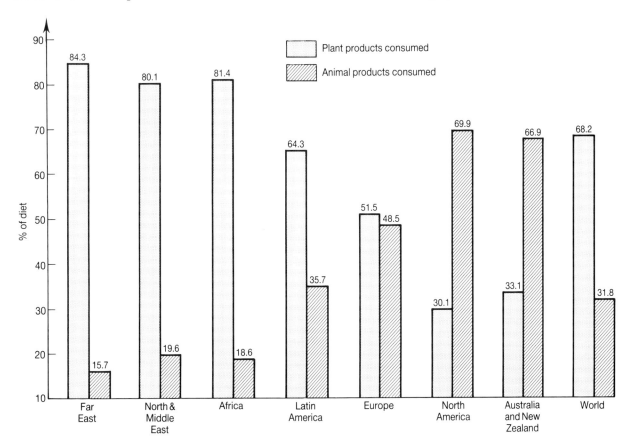

Figure 1 The contribution made by animal and plant proteins to the human diet

In the late 1980s the claim by the then Health Minister Edwina Currie that most of the eggs laid in the UK were infected with the bacterium *Salmonella*, did little to promote the sale of eggs and other products containing eggs. This claim was accompanied by a rise in the number of reported cases of salmonella food poisoning in the UK from under 9000 in 1975 to nearly 40 000 in 1988. Not all of these cases are linked to eggs though. It has been estimated that in 1988 of the 40 000 reported cases of Salmonella poisoning, only 1000 were 'egg related'. There has, however, been a drop in egg consumption of 40% over the last decade.

Since the Second World War there has been an increased awareness of the relationship between diet and health. Research has suggested that many aspects of the Western diet can be related to specific diseases. There appears to be a link between a high fat diet and heart disease, a low fibre intake and certain types of cancer and a high sugar intake and dental decay to mention but a few examples. Many people have adopted diets which contain less meat and dairy products, and more vegetable-based foods. A growing number of individuals have stopped eating meat altogether and turned vegetarian.

1.2.5 Vegetarianism

Over one million people in the UK are committed vegetarians. Many people convert to a meat-free diet on the grounds of animal welfare and the allegations of animal cruelty during production and slaughter. There are however other arguments for and against the vegetarian diet.

The average adult only requires about 50 g of protein a day. Most people living in developed nations eat much more than this. Meat is not only a source of protein but also a source of fat. By cutting meat out of our diets it is claimed that we are able to reduce our fat intake by 25%, thus reducing the risk of heart disease. However, the quality of the proteins found in vegetable sources is lower than that found in animal sources and so it is important for vegetarians to select their food carefully. Vegetarians who do not eat eggs and meat will be deficient in the vitamin B_{12} and so may require supplements. There are also arguments for vegetarianism based on the flow of energy through the environment. These will be discussed in a later section.

Table 1.4 provides a summary of some of the more poignant arguments for and against the vegetarian way of life.

Table 1.4 *The pros and cons of the vegetarian diet*

FOR VEGETARIANISM	AGAINST VEGETARIANISM
The breeding and slaughter of animals for food is cruel and degrading.	Unless animals are kept for food they will die out. If we still require animal products such as milk and eggs, all male animals will be surplus to requirement and so would still need to be slaughtered.
Animals suffer when in transit and during the process of slaughter. This is unnecessary.	There are strict regulations to control the handling and transportation of domestic animals which limit the distress which they suffer. Introduction of the humane killer has reduced distress at the slaughterhouse.
Animals are as much a part of the world as we are. We should not destroy our fellow creatures.	Most animals kill other animals to survive, being part of a predator-prey relationship.
The vegetarian diet gives as much nourishment as a meat-based diet. Food sources contain a balanced mixture of both carbohydrate and protein. The meat-eater must add starchy foods to his diet.	Meat contains high quality proteins which are made up of all the essential amino acids. Plant proteins tend to have a much lower protein quality score.
Animal proteins are associated with a high fat content and a lower fibre content. Both of these factors have been linked to health risks.	An exclusively vegetarian diet can lead to intestinal problems, especially if food is undercooked. The differences in life expectancy between a meat-eater and a non-meat-eater are very small.
Energy is lost at each stage of the food chain and so it is 'wasteful' to feed plants to animals to convert into meat.	Plant material is much more difficult for the human digestive system to break down and so nutrients are less likely to be assimilated from plant sources.

1.3 THE DOMESTICATION OF ANIMALS

1.3.1 What is domestication?

Most animal products used by humans come from **domesticated** species, for example cattle, sheep, chickens. The domestication of an animal species involves the taming of wild animals so that they are able to live in captivity. However, for an animal species to be completely domesticated we must also have knowledge of, and be able to control, its life cycle. The manipulation of the natural life cycle of an animal allows us to maximise the yield of products obtained from it. Animal breeders must be able to control the animal's reproductive cycles and breeding if maximum productivity is to be obtained.

1.3.2 Which species are suitable for domestication?

Only a small proportion of the total number of animal species known have been domesticated. Certain species are more suitable for domestication than others and some are totally unsuitable. There are a variety of reasons for this. In many cases the natural characteristics of the animal determine whether or not it is a suitable candidate. This is illustrated in table 1.5. However, tradition also plays an important part in our choice of domestic animal. Different cultures around the world choose to keep different types of animals for food production. In many areas of Europe, for example, horses are commonly reared for meat. In the UK this practice is largely unacceptable.

Tigers are a good example of a species which is unsuitable for domestication. They live solitary lives, and are strongly territorial, marking out their range with strong-smelling urine to deter other animals from entering. Like many other animals they breed in the spring. The male vists the nearest female on heat and once mating is completed, he returns to his own range having no more to do with his offspring. After birth the mother cares for her young until they are 2–3 years old. If you were to put a tick against each criterion in table 1.5 fulfilled by tigers, you would find many more under the heading 'unsuitable' than elsewhere.

A male tiger

Table 1.5 Characters for domestication

SUITABLE	UNSUITABLE
Live in large social groups e.g. flocks, herds.	Live in small families or live alone.
Group structure is based on leadership.	Group structure is based on territory.
Males and females live together in the wild.	Males live in separate groups.
Individuals will mate with any other individual of the same species.	Individuals pair and mate for life.
Female accepts young soon after birth. Bonds with her own offspring only.	Young accepted by the group due to species characteristics.
Undisturbed by humans or changes in the environment.	Very wary of humans or changes in the environment.
Able to feed on a range of foodstuffs.	Requires a specific diet.
Adapts to a wide range of habitats.	Requires a very specific habitat.

(Hale, EB, (1962) 'Domestication and the Evolution of Behaviour', *Animals for Man*. JC Bowman/Edward Arnold (1977))

Sheep, on the other hand, are eminently suitable for domestication. In the wild, they live in large groups or flocks. They are able to graze efficiently on very rough grassland as well as feed on stored forage like hay inside if necessary. They do not form permanent pairings and so will mate with any member of the opposite sex within the flock. The female cares for her young after they are born, and a very strong bond develops between a mother and her offspring. Sheep fulfil every 'suitable' criterion in table 1.5.

Cave drawings in Utah, USA show humans hunting for food

Flock of sheep being rounded up by sheepdogs

1.3.3 The history of domestication

The domestication of animals occurred as part of the process of agricultural developments which started at the end of the Stone Age. Cave drawings done before this time show that people obtained their food mainly by hunting wild animals and gathering wild plants.

Around 10 000 years ago, people began to farm, obtaining food largely from cultivated crops and livestock. The first animal to be domesticated was the dog, followed by the reindeer, goat and sheep.

The process of domestication began when humans started to capture wild animals and use them for transport and traction, and to provide food and fibres such as wool. Eventually animals with desirable characteristics were selected, captured, tamed and allowed to mate at random in captivity. In many cultures, the 'best animals' were often offered as sacrifices and so this led to a 'dilution' of characteristics. The final step on the pathway to domestication was the selective breeding of animals in captivity. This enabled the producer to select the characteristics most suited and mate animals possessing them. This process took place over several thousands of years and is summarised in figure 3.

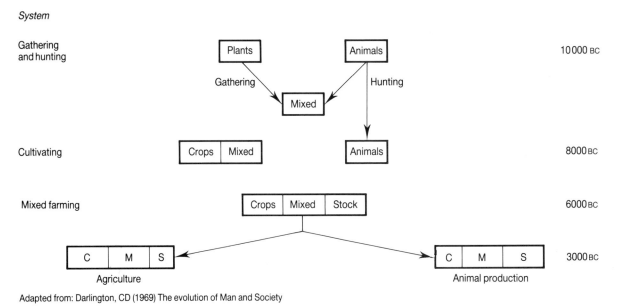

Adapted from: Darlington, CD (1969) The evolution of Man and Society

Figure 2 The history of domestication

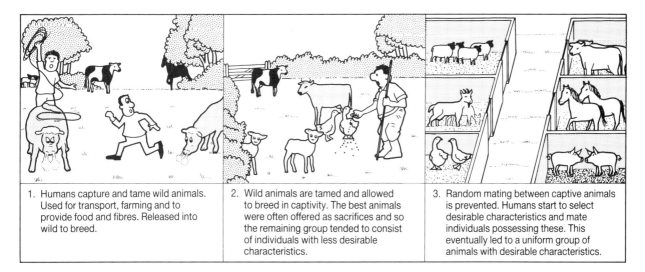

1. Humans capture and tame wild animals. Used for transport, farming and to provide food and fibres. Released into wild to breed.

2. Wild animals are tamed and allowed to breed in captivity. The best animals were often offered as sacrifices and so the remaining group tended to consist of individuals with less desirable characteristics.

3. Random mating between captive animals is prevented. Humans start to select desirable characteristics and mate individuals possessing these. This eventually led to a uniform group of animals with desirable characteristics.

Figure 3 Events leading to domestication

1.3.4 The domestication of animals and the future

While human populations remained relatively small, it was possible for them to obtain enough food from the wild. But, as the world's population began to grow in size, excessive strain was placed upon the natural environment in which communities lived. One of the reasons that domestication of animals and cultivation of crops began in many regions, was to increase food production to feed the ever-growing community.

Today, with the global population continuing to increase, it is estimated that 20% of the world's people are malnourished. It may be possible to increase food production by the domestication of species like the eland (a large ox-like antelope native to Africa) or the red deer. In this way we could increase the productivity of land which is unsuitable for the rearing of existing domestic species like sheep and cattle. The domestication of some fish species and their rearing in fish-farms has already improved the food-producing efficiency of lakes and rivers in many areas of the world. We shall look further into fish-farming in Chapter 8.

1.4 ENERGETICS AND ANIMAL PRODUCTION

The sun provides the Earth with a constant supply of radiation. Plants utilise the visible wavelengths of light in several metabolic processes, the most important of which is **photosynthesis.** During this process, solar energy is used to produce **dry matter** (or food) from carbon dioxide and water.

This means that plants form the first **trophic** level of all food chains. A food chain describes the way in which energy and food is transferred from organism to organism. Some of the dry matter produced by the plant is used up as a substrate for respiration. The rest is available as food for herbivores and omnivores. Herbivorous organisms form the second trophic level of the food chain and, again, these organisms metabolise some of the food eaten, storing some as muscle to be passed on as meat for organisms at the next level. Food and energy will be passed on in this way until it reaches a consumer at the top of the food chain, for example humans. At each level of the food chain about 90% of the energy transferred is lost through respiration and to decomposing organisms. Only 10% is available as food for the next level. This means that it is more economical to eat organisms from the lower trophic levels as shown in figure 4.

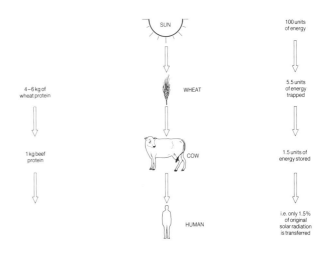

Figure 4 The approximate fate of energy and protein within a simple food chain

A square metre of grassland in the UK will receive about 1 046 700 kJ of solar energy a year. Only 21 436 kJ of this will be used for growth. If this grassland is grazed by beef cattle, only part of this dry matter will be available to the cattle. Much is lost to decomposer organisms living in the soil, and other herbivores, such as rabbits. Of the 1 046 700 kJ of energy originally intercepted, only 125 kJ will be used by the cattle for growth. This is less than 0.012%! This is outlined in figure 5.

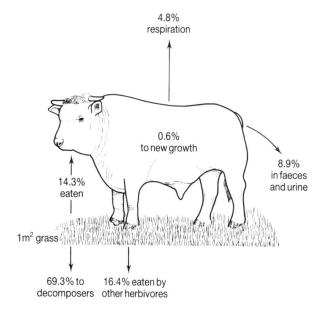

Figure 5 The fate of one year's growth of one square metre of grass grazed by cattle

Table 1.6 below compares the amount of energy and protein available to human beings from a number of food chains. This data again serves to emphasise the increased energy yield available when food chains do not involve animals. It is important to remember, however, that whilst plant-based food chains appear to yield more energy and protein, the quality and digestibility of nutrients from these sources may be lower.

1.4.1 Raising cattle on rainforest land

Tropical rainforests are one of the most productive areas of the world. Table 1.7 compares the biological productivity of the rainforest with land used for growing crops. It can be seen that the productivity of the rainforest is only exceeded by areas of sugar-cane plantation which are highly irrigated. However, when areas of rainforest are cleared and used for agriculture the productivity falls dramatically.

Table 1.7 Biological productivity of land

AREA	AVERAGE PRODUCTIVITY (tonnes of plant material/ ha/pa)
Rainforest	90
Irrigated sugar-cane	140
Potatoes	30
Wheat	15

Under natural conditions, leaf litter produced by the rich vegetation of the rainforest falls onto the soil surface and decomposes rapidly. This replaces nutrients used by the plants for growth. The high content of organic matter also helps maintain the soil structure, as does the presence of large tree roots which help prevent erosion.

When areas of tropical rainforest are cleared for agriculture, the trees are cut down and burned to expose bare soil. Tropical soils tend to weather rapidly due to the higher temperature and humidity. This means that soils allow water to percolate through rapidly. Although tropical soils are initially suitable for crop or grassland production, continual heavy rain leaches nutrients from the topsoil making them unavailable for plant growth. During the dry season, the soil dries out and becomes very hard.

Table 1.6 Energy and protein yields from some food chains

FOOD CHAIN	EXAMPLE	Energy yield to humans ($kJ \times 10^3 \ ha^{-1}$)	Protein yield of food to humans ($kgha^{-1}yr^{-1}$)
Crop → human	Monoculture of wheat	7 800–11 000	42
Crop → livestock → human	Barley-fed beef	745–1423	10–15
Intensive grassland → livestock → human	Intensive beef or dairy herd on grassland	Meat 339 Milk 3813	Meat 4 Milk 46
Grassland and crops → livestock → human	Mixed dairy farm	1356	17

(Duckham and Masefield (1970) *Farming Systems of the World*)

When heavy rain falls on this dry soil, there is a danger of erosion and soil loss.

Newly cleared tropical soils are extremely fertile due to:

- the organic matter contained within the soil due to the decomposition of leaf litter,
- the nutrients added to the soil as a result of the burning of trees.

However, after a few years the fertility level of the soil declines because nutrients are being removed by plants for growth and are also subject to leaching. Leaching makes it difficult to maintain the fertility level of the soil.

The rainforest is being destroyed at an alarming rate. The land made available by the clearance of trees is often not used to produce food for the local communities, but is bought by large companies and used to produce food for profit. In 1987, it was reported that over 200 000 km^2 of rainforest was cleared by burning in Brazil alone. Many big companies own large ranches on which cattle are reared for beef production. This system brings no money into the local community, destroys the natural rainforest habitat and, because the land is cleared by burning, increases the concentration of greenhouse gases in the atmosphere. In terms of energetics if this land resource has to be destroyed to provide food, it would make more sense to utilise the area for local crop production.

1.5 SYSTEMS OF ANIMAL PRODUCTION

In the wild, animals obtain their food, water and shelter from their natural environment. Once animals have been domesticated, their human keepers must provide these resources for them in one way or another. The degree to which humans interfere with the lifestyle of the animal they exploit varies with species and with the human requirement of the animal. At one extreme we **exploit** a natural environment by harvesting animals from it. Sea-fishing is an example of this exploitative production where the human impact on the animal's lifestyle is minimal. At the other extreme is **intensive** farming, when the whole of the animal's lifestyle is manipulated by its human keepers. An example of this may be the production of eggs on a battery farm. Between these two extremes is **extensive** or **free-range** farming, which involves an intermediate degree of human interference in the animal's way of life.

1.5.1 Exploitative animal production – sea-fishing

People have harvested sea-fish from their natural habitat for thousands of years, with little regard for the consequences. Nowadays we are more aware of the detrimental effects of overfishing on the whole ecology of the sea and, of course, the populations of the fish themselves. It is possible to estimate the annual natural production of fish in different regions of the sea and use this information to set a limit on the numbers of fish harvested in each area.

A campaign to broaden consumer tastes in fish would also help reduce the overfishing of those fish which are currently popular. Only a small proportion of fish species are used as food in most parts of the world because consumers have limited knowledge of what can be eaten. Another suggested solution for the problem of dwindling fish populations is that unwanted fish species could selectively be removed from the sea so that more food is available for the more popular fish so that their numbers can increase.

However, perhaps a more feasible solution is to raise marine species in demand for food in artificial conditions. Special hatcheries already exist which produce many popular species, such as plaice, sole, turbot and carp.

Traditional sea-fishing in the Seychelles

1.5.2 Extensive animal production

Sheep were one of the first species to be domesticated by humans. Numerous breeds of sheep now exist and between them they are adapted to a whole range of environments. As sheep are able to consume low grade plant material, high in cellulose, which many animals cannot, they are farmed in countries throughout the world.

We obtain three products from sheep – wool, milk and meat. Wild sheep shed their coat each year but this characteristic has been removed from the domestic sheep by selective breeding and so they have to have their wool shorn. Depending on age and species, a sheep can produce between 0.5 – 10 kg of wool a year. Sheep which are kept for milk require a lot more attention from their keepers as they obviously require regular milking. They can produce up to 100 litres of milk per season and this is used mainly for cheese production. Sheep which are kept for meat production are slaughtered from a few weeks old upwards. Lamb is higher in fat than most other meat, containing up to 35% fat.

Very little human labour is needed for the care of sheep (except milk sheep). Sheep obtain their food by grazing and often find their own shelter. Human interference is limited to shearing, assisting with lambing and the removal or castration of surplus males.

Sheep production

1.5.3 Intensive animal production

The domestic chicken is found in human settlements worldwide and in many cases lives under free-range conditions as a scavenger. Many live solely on household scraps and under these conditions a hen will produce 80–150 eggs a year during the laying season. The laying season is controlled by day-length. It begins when the day-length starts to increase in the spring and ends in the autumn when the day-length decreases. The hen is allowed to lay for two or three seasons and then is killed for meat. As cockerels are allowed to run with the hens, many of the eggs produced by this 'free-range' method will be fertilised. Some of these fertilised eggs are allowed to develop and hatch by the farmer, to replace those chickens killed for meat. This system of egg production has been mainly replaced in the developed world by a more intensive system.

Varieties of hen have been selectively bred which can lay 250–280 eggs a year. These strains have also been selectively bred so that they convert feed more efficiently, so egg size and weight is greater than for wild birds. Producers buy their hens as day-old chicks and keep them either on the floor of barns on litter or in cages. It is common for a producer to keep up to 1 000 000 birds in a single unit (at a density of 10 birds per 1 m^2.) The birds are given specially formulated feed and water, often automatically, and daylight and temperature are controlled. The hens are kept for one laying season (12 months), and then slaughtered for pet food.

Battery egg production

QUESTIONS

1 Britain has 18 million hectares of agricultural land. The table below shows how it is utilised.

USE	HECTARES $\times 10^6$
Growing animal fodder	2.3
Grazing land for animals	13
Growing plant food for humans	2.8

a) Represent these data visually, using the technique that you feel is most appropriate.
b) Why do you think such large areas of land are needed for animal production compared with those required for crop production?
c) List what you think would be the advantages and disadvantages of British farmers using more of their agricultural land for growing plant food for humans
   (i) to British people?
   (ii) to the people of the developing world?

2 The table below shows the approximate contribution made by various foods to the protein supply of humans in Africa, North America and worldwide.

a) Draw pie charts to show how humans obtain their protein in each region.
b) Calculate the percentage contribution to the protein content of the diet in N. America and Africa from
   (i) vegetable sources,
   (ii) animal sources.
c) What can you conclude about the relative protein contributions of animals and plants to the diets of these human populations? Suggest reasons for your conclusions.

3 Think of the reasons why people consciously choose to limit their diet in some way. Talk to others to find out their beliefs concerning food and prepare a list of reasons. Now imagine you are a person who firmly believes in one or more of these reasons. Write a short article for a magazine in which you explain your beliefs. Back up your argument with facts and counter any foreseen objections from others.

4 a) What effects do you think the overfishing of a particular species in a particular region, will have on the food chain in that region?
b) Find out the names of three species of fish in the developed world
   (i) commonly used for food,
   (ii) rarely used for food.
c) How would you go about selectively removing unwanted species of fish from the seas? What effects would you envisage this having on the ecosystem?
d) What other examples can you find of how man exploits a natural population to obtain food?

5 a) Intensively farmed hens produce up to 280 eggs a year. Egg shell is rich in calcium. What effects do you envisage that this high rate of laying will have on the hen?
b) Why do you think that intensively farmed hens are only kept for one year?
c) Why is disease control important in intensive egg production?
d) Many people violently disagree with the methods used in intensive egg production. Producers argue that such methods are essential to meet the huge demand for eggs at a reasonable price to the consumer. With a partner, discuss this issue. One of you should take the role of the egg producer and the other should argue against intensive farming. Try to bring in as much factual evidence as you can and afterwards note down in a table the main points on each side of the argument.
e) Find out as much as you can about another example of intensive animal production and write a short report about it.

REGION	CEREALS %	ROOTS AND TUBERS %	PULSES, NUTS, SEEDS %	VEGETABLES AND FRUIT %	MEAT AND FISH %	MILK AND EGGS %
North America	18	3	4	5	39	31
Africa	55	9	16	2	13	5
World	48	4	12	4	18	13

Food source	Protein	Carbohydrate	Fat	Vitamin C	Vitamin D	Vitamin B_{12}	Iron	Calcium
Animal	Meat Fish Milk Eggs Cheese		Milk Butter Cheese	Liver	Liver Fish-oil Cheese Eggs	Cheese Eggs Meat	Meat Eggs	Meat Eggs Fish Cheese Milk
Plant	Nuts Pulses Cereals	Pulses Cereals Potato Fruit	Nuts	Fruit Vegetables			Pulses Green vegetables	Green vegetables Bread Potato

6 The table above shows some of the nutrients required by humans to remain healthy. Which nutrients, if any, are likely to be deficient in

a) a person eating both plant and animal products?

b) a vegetarian who eats no meat but does eat eggs and milk products?

c) a vegan who does not eat any animal products?

d) How could a person who has consciously chosen a particular diet try to combat any nutrient deficiency that may arise?

7 Examine the figures in the table below:

SPECIES	NO. PREGNANCIES/YEAR WILD	NO. PREGNANCIES/YEAR INTENSIVE FARMING	NO. OFFSPRING/PREGNANCY WILD	NO. OFFSPRING/PREGNANCY INTENSIVE FARMING
Pig	1	2.2	6	11
Sheep	1	1	1	2
Cow	1	1	1	1

a) For each species calculate the number of young produced each year in the wild and on an intensive farm.

b) How do you think farmers are able to increase the reproduction rate of some animals if they are farmed intensively? Why is it that for some animals reproduction rate is not so strongly affected?

c) Which of the species in the table do you think would be suitable for intensive farming?

8 Of the solar energy falling on a field grazed by cattle, only a small proportion is absorbed by grass. Give the possible fates of

a) the solar energy which is not absorbed,

b) the energy in the grass that is eaten by the cattle.

BIBLIOGRAPHY

Acker, D., Cunningham, M. (1991) *Animal Science and Industry 4th Edition.* Prentice Hall.

Bowman, J.C. (1977) *Animals For Man.* (IOB Studies in Biology no.78) Edward Arnold.

Clutton-Brock, J. (1981) *Domesticated Animals.* Heinemann.

2 VARIATION IN ANIMALS

LEARNING OUTCOMES

After studying this chapter you should be able to:
- distinguish between discontinuous and continuous variation,
- construct and interpret frequency distribution curves,
- talk about the contribution of environmental and genetic factors in the genotype of an animal,
- recall the fact that characteristics can be inherited from one generation to the next through the passing on of genes which are found on the chromosomes,
- state the Laws of Inheritance and relate these to worked examples,
- distinguish between cases involving co-dominance and those showing normal dominance patterns,
- define the term **multiple genes** and **polygenes** and describe the differences between them,
- outline the process of epistasis,
- define the term heritability and calculate the heritability of characteristics from the data supplied.

2.1   TYPES OF VARIATION

Individuals within a population vary from each other. Variation may be **discontinuous** where the characteristics exhibited fall into a few sharply defined groups, or **continuous** where a whole range of differences are possible, falling between two extremes. Most individuals exhibit characteristics with values approaching the mean, giving rise to a normal distribution throughout the population as a whole.

2.1.1 Discontinuous variation

Discontinuous variation is common between species of animals and between breeds of the same species. For example, the photos below show how the shape of the comb on the head of chickens varies with breed. The single comb is characteristic of many breeds including the Leghorn and the Rhode Island Red, the rose comb is found in the Wyandotte breed. The pea-comb is found in Indian Game and Brahma breeds and the walnut comb is a characteristic of the Malay. Comb shape is an example of discontinuous variation because each breed will have one of these distinct comb types. There are no intermediate forms. Characteristics which vary in this way are used when developing identification keys.

Comb types in chickens

2.1.2 Continuous variation

Many characteristics exhibited by individuals vary continuously. Characteristics which vary continuously are far more common.

Table 2.1 shows an example of continuous variation in pigs. This example examines the number of males produced in each litter of eight piglets born. It can be seen that only one of the 250 litters investigated contained no males and only one contained eight males. These are the two extreme conditions. Between these two extremes any number of males within the litter are possible. If a histogram is drawn of the results as shown in figure 6 a normal distribution is produced with the majority of litters containing four males (the mean number).

Table 2.1   Number of male piglets per litter of eight

CLASS INTERVAL	TALLY	TOTAL
0–0.9	ı	1
1–1.9	ᕼᕼ ııı	8
2–2.9	ᕼᕼ ᕼᕼ ᕼᕼ ᕼᕼ ᕼᕼ ᕼᕼ ıı	32
3–3.9	ᕼᕼ ᕼᕼ ᕼᕼ ᕼᕼ ᕼᕼ ᕼᕼ ᕼᕼ ᕼᕼ ᕼᕼ ᕼᕼ ᕼᕼ ı	56
4–4.9	ᕼᕼ ᕼᕼ ᕼᕼ ᕼᕼ ᕼᕼ ᕼᕼ ᕼᕼ ᕼᕼ ᕼᕼ ᕼᕼ ᕼᕼ ᕼᕼ ᕼᕼ ᕼᕼ	70
5–5.9	ᕼᕼ ᕼᕼ ᕼᕼ ᕼᕼ ᕼᕼ ᕼᕼ ᕼᕼ ᕼᕼ ᕼᕼ ᕼᕼ	50
6–6.9	ᕼᕼ ᕼᕼ ᕼᕼ ᕼᕼ ᕼᕼ ııı	28
7–7.9	ᕼᕼ ᕼᕼ	10
8–8.9	ı	1

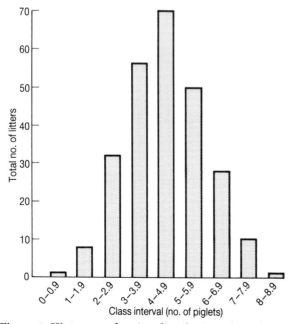

Figure 6 Histogram showing distribution of number of male piglets in litters of eight

2.2 CAUSES OF NATURAL VARIATION

Animals within a population will vary naturally from one another. In situations where animals are being produced for meat or other products this variation is often a problem to the producer. To ensure that the final product is fairly standard and to maximise efficiency, it is often desirable for the producer to try to minimise variation within a herd or flock. Variation has two components:

(i) the **environmental** component which may include such factors as level of nutrition, quality of housing, temperature of surroundings, attack by pests and diseases. These factors can be manipulated by the producer and good husbandry techniques enable him to manipulate these environmental influences to maximise the potential of his animals,

(ii) the **genetic** component which is due to inherited differences.

Figure 7 below shows the effect of these two components on the weight of 400 day-old beef cattle.

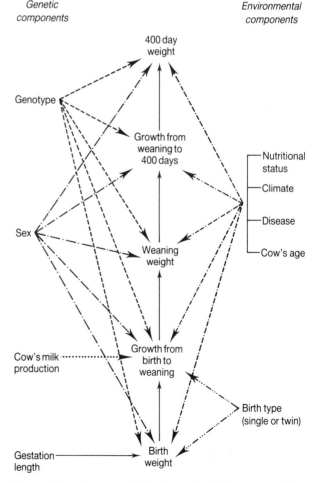

Figure 7 Environmental and genetic influences on the weight of 400 day-old beef cattle

The diagram shows how a beef calf grows from birth, through weaning to its 400 day weight. Its birth weight will be affected by mainly genetic factors such as its sex and the number of calves produced by the cow (i.e. twin or single). As the calf grows it is fed on its mother's milk. The amount of milk produced (yield) is determined, in part, by the genetic make up of the mother. Other factors such as the age of the cow and its nutritional status will also have an influence. The growth rate at this stage will also depend on whether the cow is feeding one or two calves. In cattle production, the calf is left to feed from its mother for a few days only. It is then reared on formula milk. This would remove any variation in growth rate due to the composition or quantity of milk the cow produces. After weaning, the continued growth of the calf will depend mainly on environmental influences which can be controlled by the producer. These will include, type and amount of food eaten, climate, and diseases. The main genetic influences on growth rate will be breed and sex, as males will grow more quickly than females.

2.3   GENE INTERACTION

The genetic component of variation is due to the inheritance of characteristics from parents. Characteristics are passed on from one generation to the next by the chromosomes. Chromosomes are found in the nucleus of all cells and are made up of genes which are the units of inheritance. The structure and function of chromosomes and genes are discussed fully in many standard A level Biology texts including *Genetics and Evolution* by Michael Carter from the **Focus on Biology** series.

The way in which characteristics are inherited was discovered by Gregor Mendel in 1865. His work lead to the formulation of the Laws of Inheritance. Mendel's work is well documented in A level texts and so will be dealt with only briefly here.

2.3.1   The Laws of Inheritance

The First Law of Inheritance: The Law of Segregation.
    The characteristics of an organism are determined by internal factors (genes) which occur in pairs. Only one pair of such factors can be represented in a single gamete.

The Second Law of Inheritance: The Law of Independent Assortment.
    Each member of a gene pair may combine randomly with either of another pair.

In general, each characteristic is determined by a pair of genes or **alleles**. Each gene has two forms – one **dominant** to the other, and this form is always expressed if it is contained within the genetic make up of the animal (the **genotype**). The other gene form is said to be **recessive**.

For example, Holstein-Friesian cattle may be black-and-white or red-and-white in colour. Coat colour is controlled by a pair of genes which can exist in two forms:
    (i) dominant form (represented by B) which produces a black coat,
    (ii) recessive form (represented by b) which produces a red coat.

Gamete cells will only contain one of these two gene forms as they only contain half the genetic information found in normal cells. During fertilisation two haploid gametes combine to form a new diploid individual. The zygote formed will now contain a pair of these genes, the combination of which will determine coat colour in the animal produced. Animals with the genetic make up (genotype) BB and Bb will make the coat black and bb will produce a red coat. The actual colour shown is called the **phenotype** of the animal. BB and bb are called **homozygous** individuals because both gene forms are the same, Bb are called **heterozygotes**. The inheritance of coat colour follows the laws of inheritance and is summarised in figure 8.

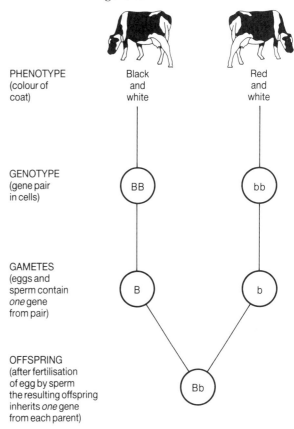

Figure 8 The inheritance of coat colour in Holstein-Friesian cattle

2.3.2 Co-dominance

Not all pairs of major genes exhibit full dominance. Consider the inheritance of coat colour in Shorthorn cattle. In this breed coat colour is controlled by two genes R (red) and r (white). Inheritance is similar to that for Holstein-Friesians, except that R is not dominant to r and the heterozygote Rr will produce cattle of a 'roan' colour i.e. their coats contain a mixture of red and white hairs.

Roan Shorthorn

2.3.3 Multiple genes

In reality, many characteristics are controlled by **multiple genes**. A multiple gene has more than two forms. Within a population it is possible to find examples of all the gene forms, however an individual is still only able to carry one pair of the forms. Characteristics which vary discontinuously and produce distinct groups of individuals, are likely to be controlled by multiple genes. Multiple genes sometimes have a cumulative effect on the phenotype and sometimes show dominance, but not always.

Animal and human blood groups are good examples of multiple genes. The inheritance of blood groups in domestic farm animals is very complex, and often controlled by many gene pairs (for example more than 20 in poultry). Human blood groups are slightly less complicated and are controlled by three genes, A, B and O. A and B are **co-dominant** to each other and dominant to O. An individual can only carry two of these genes and so the following genotypes are possible.

Table 2.2   Genotypes of human blood groups

Blood Group	Genotype
A	AA, AO
B	BB, BO
AB	AB
O	OO

2.3.4 Polygenes

Most important production traits in farm animals (such as milk yield, growth rate, carcass composition) are controlled by many pairs of genes rather than one single pair. These **polygenes** are collections of large numbers of genes, each of which are of relatively minor importance individually, but which collectively have cumulative importance on the value for one particular trait. They are additive in their effects and so control examples of continuous variation. As with characteristics controlled by co-dominance, polygenes do not exhibit normal dominant/recessive inheritance but are cumulative in effect.

Milk yield in dairy cows is controlled by polygenes. Let us assume that 3 pairs of genes – Aa, Bb, Cc – are responsible for controlling milk yield, although in reality there will be many more genes than this involved. Assume that the combination aabbcc gives a lactation yield of 4000 litres and that each dominant gene adds a further 400 litres to the yield. Thus the other extreme genotype (AABBCC) will have a yield of 6400 litres (i.e. $4000 + (6 \times 400)$).

Figure 9 shows the effects of crossing individuals with these two extreme genotypes. The individuals produced have the genotype AaBbCc and will yield 5200 litres of milk.

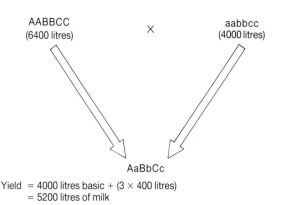

Figure 9 Crossing AABBCC with aabbcc

If these individuals (AaBbCc) are allowed to interbreed, a whole range of offspring are produced. Table 2.3 shows the effect of this cross and the milk yields expected from the potential offspring. If these milk yields are plotted onto a histogram a normal distribution is obtained, with most individuals yielding 5200 litres of milk. This is the mean.

Table 2.3 Expected milk yields (litres) from interbred offspring

GAMETES	ABC	ABc	AbC	aBC	Abc	aBc	abC	abc
ABC	6400	6000	6000	6000	5600	5600	5600	5200
ABc	6000	5600	5600	5600	5200	5200	5200	4800
AbC	6000	5600	5600	5600	5200	5200	5200	4800
aBC	6000	5600	5600	5600	5200	5200	5200	4800
Abc	5600	5200	5200	5200	4800	4800	4800	4400
aBc	5600	5200	5200	5200	4800	4800	4800	4400
abC	5600	5200	5200	5200	4800	4800	4800	4400
abc	5200	4800	4800	4800	4400	4400	4400	4000

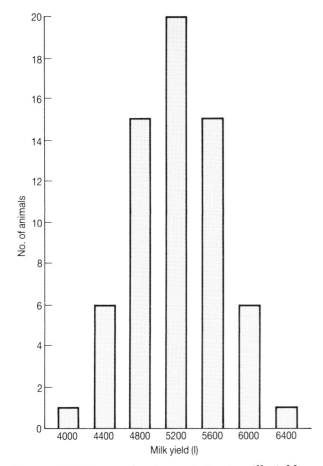

Figure 10 Histogram showing variation in milk yield due to polygenetic inheritance

2.3.5 Epistasis

In the examples considered so far, characteristics have been inherited from one generation to the next because the gene controlling each has existed as one of a pair. Each pair has a dominant and a recessive form, and the dominant form is always expressed when present. However, in some cases of inheritance, the expression of one gene is controlled by the presence or absence of another gene which may even exist on a different chromosome. This is called **epistasis**.

Epistasis occurs with feather colour in poultry. Varieties of domestic fowl (*Gallus domesticus*) can be used to illustrate this. Feather colour is controlled by two interactive pairs of genes:
  (i) the colour genes R = red, r = white,
  (ii) an epistasic gene which inhibits the expression of the colour gene, I = no colour, i = colour.

If birds are to have red feathers their genotype must contain not only the dominant red colour gene (R) but the dominant epistatic gene (I) must be absent.

Table 2.4 Epistatic control of feather colour in poultry

PHENOTYPE	GENOTYPES
Red feathers	RRii, Rrii
White feathers	RRII, RRIi, RrII, RrIi, rrii

2.4 HERITABILITY

Variation between individuals is influenced by the environment in which they live and the characteristics which they inherit from their parents. Each characteristic exhibited by an individual may have both environmental and genetic influences. If an animal producer wishes to breed an animal for one particular characteristic he must know what proportion of that characteristic is influenced by genetics. If inheritance has a large effect on the desired characteristic then the producer must select animals for mating carefully. If, however, the environment has a larger influence, the producer must pay more attention to husbandry techniques. The genetic influence on a characteristic is called the **heritability**.

Heritability can be expressed as a proportion ranging from 0–10 or a percentage (0–100%). If the heritability of a characteristic is above 50% then inheritance plays a large part in the expression of the characteristic. Some examples are shown in table 2.5.

Table 2.5 Heritability of selected characteristics

SPECIES	CHARACTERISTIC	% HERITABILITY
Cattle	Milk yield	43
	Birth weight	49
	Slaughter weight	85
Sheep	Birth weight	33
	Fleece weight	17
	Multiple birth	4
Chicken	Egg production	21
	Egg weight	60
Pig	Litter size	4
	Carcass length	24
	Leg length	48

2.4.1 Calculating heritability

The heritability of a particular characteristic from one generation to the next can be calculated if the variance is known. **Variance** describes the way in which a characteristic e.g. milk yield, varies around the average value or mean. If the milk yields of a herd of cows are recorded a histogram can be produced. This will show that yields vary between two extremes with most individuals in the herd having yields around the mean. Variance is calculated using the equation below:

$$\text{Variance} = \frac{\Sigma n\,(x - \bar{x})^2}{n - 1}$$

Σ = sum of
n = number of individuals
$\bar{x}$ = mean
$x - \bar{x}$ difference between individual and mean

To calculate the heritability of milk yields the variance due to environment and genetics must be calculated.

Total Variance = Genetic Variance + Environmental Variance
(Vt)   =   (Vg)   +   (Ve)

Heritability is the proportion of total variance which is due to inheritance or genetics and not the environment.

$$\text{Heritability (H)} = \frac{Vg}{Vt}$$

Now try Investigation 1 The Heritability of Bristles in *Drosophila* in *Animal Science in Action Investigations*.

QUESTIONS

1 Consider the coat colour of Holstein-Friesian dairy cattle. Two basic coat colours exist, black-and-white and red-and-white. The black gene (B) is dominant to the red gene (b).
   a) What possible genotypes can exist? In each case also state their phenotype.
   b) What colour cattle would be obtained if
      (i) a pure breeding red-and-white female was crossed with a pure breeding black-and-white male?
      (ii) two offspring from the cross in (i) were interbred?
   In each case state the genotypes and phenotypes of all individuals.

2 Hereford cattle are produced mainly for their meat, rather than for their milk. The Hereford breed has a distinctive white face. This characteristic is controlled by a dominant gene. Hereford bulls are often crossed with other breeds (especially dairy cattle) to produce cross breeds suitable for veal and beef production. These cross breeds will have white faces. Explain why you think this dominant 'white face' gene is useful to animal producers.

3 Breeds of poultry used for meat production have either yellow or white skins. White skins are produced by a dominant gene (W) and yellow ones by the recessive gene (w). In the USA the consumer prefers yellow-skinned birds, whilst in the UK there is a preference for those with white skins.

   In 1962 strains of poultry were imported into the UK from the USA which had better growth rates. However these birds also had yellow skins. The UK breeders, therefore, set about transferring the white gene from the native bird into the faster-growing imported bird. The following cross was carried out:
   (i) Pure breeding white birds (WW) were crossed with pure breeding yellow birds (ww). The offspring had the genotype Ww and were all white-skinned.
   (ii) These birds were interbred producing the following genotypes: WW (white), Ww (white), ww (yellow).

   The yellow-skinned birds could be removed leaving a large number of white-skinned birds. These birds had successfully incorporated the white gene into their phenotype without losing any of the other advantages. However if these birds were to continue to produce only white-skinned offspring individuals must be homozygous for this characteristic and have no recessive (w) gene in their genotype.

a) Find out how breeders test animals and plants in order to find their genotype when this type of problem occurs.

b) Show how these methods could be used to discover the genotype of the white birds in this example.

4 a) Define the term co-dominance.

b) Coat colour in Shorthorn cattle is controlled by two genes R (red) and r (white). These two genes exhibit co-dominance and the intermediate colour is known as roan. What do you think would happen in the following cases:

(i) A red bull is crossed with a white cow?

(ii) The offspring of this cross are interbred?

c) How might co-dominance increase variation within a population? When might this be an advantage in nature and in the commercial production of animals?

5 The diagram below shows comb shapes in chickens. Two different allelic pairs of genes R,r and P,p interact in affecting comb shape. The genotype rrpp gives a single comb, R–P– gives a walnut comb, rrP- gives a pea comb and R–pp gives a rose comb. The dash (-) indicates the presence of either the dominant or recessive gene.

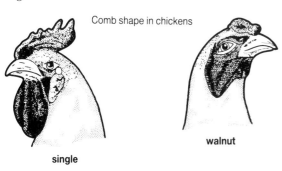

Comb shape in chickens

single

walnut

pea

rose

Comb shapes in chickens

a) Explain what is meant by allelic pairs of genes.

b) What comb shapes will appear, and in what proportions in the F_1 and in the F_2 generations, if single-combed chickens are crossed with true-breeding walnut-combed ones? Show how you derive your answer.

c) List all those genotypes which will produce a walnut-comb phenotype.

d) A walnut-combed chicken is crossed with a rose-combed one. All the progeny are walnut-combed. List all the possible genotypes of the parents. Show how you derive your answer.

e) What are the genotypes of the parents in a walnut-combed × rose-combed mating from which the progeny are ⅜ rose-combed, ⅜ walnut-combed, ⅛ single-combed and ⅛ pea-combed? Show how you derive your answer.

UCLES 1992

6 Epistasis occurs with the feather colour of poultry. The colour of feathers is controlled by two genes:

(i) The colour gene R (red) and r (white),

(ii) The epistatic gene I (inhibits colour) and i (allows colour gene to be expressed if present).

Consider a cross between two breeds of poultry shown below:

BREED	GENOTYPE	PHENOTYPE
White Wyandotte	rrii	white
White Leghorn	RRII	white

a) Why are White Leghorn chickens white when they contain the dominant colour gene in their genotype?

b) What do you think would be the outcome of crossing birds from these two breeds?

c) Epistasis also controls comb shape in domestic fowl. Find out how this gene system operates.

7 Data obtained from identical twins on the heritability of characteristics in cattle is usually much higher than estimates made using parent-offspring resemblances. What factors might account for the differences?

8 The diagram below shows variation in the butterfat content in the milk of two breeds of cattle, Red Dane and Jersey, and in a cross between them. Red Dane and Jersey cattle are homozygous for genes affecting butterfat content.

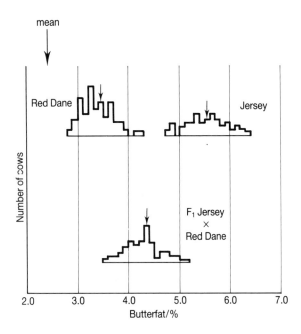

A graph to show the butterfat content of milk from three breeds of cow

a) What is the evidence that variation in butterfat content in the parent breeds may be due to:
    (i) genetic differences?
    (ii) environmental differences?

b) (i) What do you understand by the term polygenetic inheritance?
    (ii) How do the results for the F_1 generation suggest that butterfat content of milk shows polygenetic inheritance?

AEB 1992

BIBLIOGRAPHY

Brown, T.A. (1989) *Genetics, A Molecular Approach.* Van Nostrand Reinhold.

Carter, M. (1992) *Genetics and Evolution.* (Focus on Biology). Hodder and Stoughton.

Willis, M.B. (1991) *Dalton's Introduction to Practical Animal Breeding 3rd Ed.* Blackwell Scientific Publications.

3 BREEDING BETTER ANIMALS

LEARNING OUTCOMES

After studying this chapter you should be able to:
- explain how traits are selected for inclusion in a breeding programme,
- define the term **breed**,
- distinguish between inbreeding and outbreeding and evaluate the usefulness of each within a breeding programme,
- talk about the modern methods such as artificial insemination, embryo and gene transfer, used by the breeder to improve the animals produced.

3.1 DO WE NEED BETTER ANIMALS?

Ever since human beings started to domesticate animals, we have tried to improve their characteristics or traits so that they are more efficient producers. Traditionally this would be done by mating together animals which showed desirable characteristics, however, recently new scientific developments, such as embryo transfer and artificial insemination (AI), have enabled animal breeding to become more accurate and more economic.

3.2 SELECTING TRAITS

Animals have been selected for domestication by human beings because of certain characteristics which they exhibit. For example, the domestication of cattle may have taken place because of their ability to produce milk, or because of the meat they produce, or for a number of other characteristics which they possess. Some examples of desirable traits in common farm species are shown in table 3.1.

Table 3.1   Traits in farm animals

TYPE OF TRAIT	EXAMPLE
Fitness	Litter size
	Offspring survival rate
Production	Growth rate
	Milk yield
Quality	Carcass composition
	Butterfat percentage in milk
Type	Coat colour
	Physical appearance
Behaviour	Docility
	Response to stress

3.2.1 Selecting for breeding performance

If a breeder wants to produce better animals, it is important to decide which characteristics are the most important for inclusion in a breeding programme. Suitable traits must, of course, have a high heritability (see Chapter 2), and usually have the greatest economic value. For example, in the UK sheep are mainly kept for meat production and so growth rate (pre- and post-weaning gain) and the percentage of lean meat on the carcass, may be the two most important traits in a breeding programme. However, in Australia, sheep are mainly kept for their wool and so traits such as fleece quality and size will become more economically important.

The more traits included in a breeding programme, the more complex it will become and signs of improvement in the flock or herd will be slower. It is, therefore, important to prioritise the traits that you wish to improve and not attempt to attain success in all desired traits at once.

3.2.2 Heritability and trait selection

Heritability describes how genetics influence characteristics. It is expressed on a scale of 0–100% and if it has a high score it means that particular characteristic is influenced greatly by genetics and not by the environment (see Chapter 2). The heritability of some traits in domestic animals are shown in table 3.2.

If a trait is highly heritable, selection is often effective. Breeders of dairy cattle, for example, can be fairly sure that cows which produce milk with a high butterfat content are likely to pass on that trait to their offspring. If this characteristic is also of economic importance breeders are likely to pay well for a cow which shows this trait. If, however, heritability is low, improvement by selection is slow and often uneconomic.

Table 3.2 Heritability of traits in farm animals

SPECIES	TRAIT	% HERITABILITY
Sheep	Multiple births	4
	Post-weaning gain	60
	Fleece weight	17
	% lean meat	35
Beef cattle	Numbers born	5
	Post-weaning gain	50
	% lean meat	40
	Meat tenderness	60
Dairy cattle	Milk yield	43
	% butterfat	50
Pig	Litter size	4
	Post-weaning gain	25
	% lean meat	35
Poultry	Hatchability of fertile eggs	10
	Egg production	21

3.3 NATURAL MATING AND BREEDING SYSTEMS

Once a breeder has selected the traits which he wishes to improve in his animals, he is able to select which animals to breed from. He must now decide whether to mate these animals with animals from the same breed (inbreeding or pure breeding) or from a different breed (outbreeding or cross-breeding).

3.3.1 What is a breed?

'Breed' is a relatively recent term used to describe a group of animals within a species which have certain characteristics in common. Most of these traits are physical ones, such as colour. Different breeds of animals have developed either from the physical isolation of some individuals leading to selection and the disappearance of some characteristics, or by cross-breeding. (The term 'variety' tends to be used in place of breed in the plant kingdom.)

Sheep all belong to the same species but many breeds exist

3.4 INBREEDING

Inbreeding, or pure breeding, involves the mating of an animal with another of the same breed. Animals from the same breed are more closely related than animals of different breeds and the same species, and so they are likely to be genetically more similar. Inbreeding is used by livestock producers in order to concentrate genes controlling desirable characteristics so that superior offspring are produced.

All individuals of the same breed are related to some extent — the closer the relationship between breeding animals the more similar their genotypes. Inbreeding is used extensively when establishing new breeds in order to produce individuals of very similar genotypes and to select out individuals which possess distinctly different characteristics. It does however carry risks — the problems associated with this practice include inbreeding depression and the concentration of lethal genes.

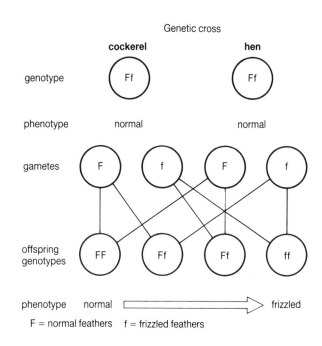

Figure 11 The inheritance of frizzled feathers

3.4.1 Inbreeding depression

Animals produced by continual inbreeding from generation to generation tend to gradually show a decrease in performance with respect to some traits. This is called **inbreeding depression.** It is caused because while inbreeding aims to concentrate desirable characteristics, it may also concentrate undesirable ones. The types of trait which are most often affected are those concerned with factors such as fertility and growth rate. The resulting offspring will be less vigorous and, therefore, less able to survive.

3.4.2 Lethal genes

Some genes act in such a way as to cause an individual to die or be severely deformed. These **lethal genes** are often recessive and so are unlikely to show in the phenotype of an individual as they will be masked by the dominant gene. However, the mating of closely related animals during inbreeding increases the likelihood of these lethal genes coming together. This is shown in figure 11.

Examples of abnormal conditions caused by lethal genes include the conditions known as 'frizzled feathers' in poultry and 'amputate' in Friesian cattle. 'Frizzled feathers' is the name given to the condition where the feathers fail to develop completely. This means that the bird is unable to keep warm and so will die from heat loss. Calves suffering from 'amputate' are born dead, with no feet and a parrot-like muzzle.

The breeder must be able to balance the effects of lethal genes, and be prepared to cull abnormal or unsuitable animals, if inbreeding is to be successful.

3.4.3 Outcrossing

Outcrossing is similar to inbreeding as it involves mating animals from the same breed. If inbreeding depression has led to the loss of a particular trait within a herd or flock, a breeder may mate the animals with a member of the same breed but from a different herd or flock which show the missing trait. If the animals are related, but not closely (unrelated for 5–6 generations) inbreeding depression is less likely. The purpose of outcrossing is to reintroduce desired traits into the phenotype.

Interspecies cross of a horse with a donkey produces a mule

Interspecies cross of a sheep and a goat produces a gheep

3.5 OUTBREEDING

Outbreeding involves the mating of an animal with another either from another species or from another breed.

3.5.1 Interspecies crosses

Mating animals from one species with those of another is rarely successful. If animals have different chromosome numbers embryos often do not survive, and if they do, they are usually sterile. The most common interspecies cross (or hybrid) is between the donkey and the horse, producing a mule. Other cross species animals include the beefalo (buffalo and cattle) and the gheep (sheep and goat).

3.5.2 Cross-breeding

Cross-breeding involves the mating of an animal with another of the same species but of a different breed. Cross bred animals tend to be more vigorous than pure bred animals, showing increased survival and growth rates along with increased productivity.

Cross-breeding causes **hybrid vigour.** Hybrid vigour occurs because most desirable traits are controlled by dominant genes, whereas detrimental traits are controlled by recessive ones. Cross-breeding causes the genes from two breeds to become combined. This is likely to increase the number of traits controlled heterozygotically (i.e. by one dominant and one recessive gene) and so the offspring are likely to show the desirable characteristic controlled by the dominant gene in the phenotype. Table 3.3 shows a simplified example of this.

Table 3.3   Simplified example of how cross-breeding increases hybrid vigour

	BREED X	BREED Y	CROSS BRED OFFSPRING
Number of pigs per litter	16 (AA)	4 (aa)	16 (Aa)
% surviving after 10 days	25 (bb)	100 (BB)	100 (Bb)
Number of piglets alive at 10 days	4	4	16

Imagine two breeds of pigs exist; breed X which produces about 16 piglets per litter but which only have a survival rate of 25% at 10 days (only 4 of the 16 survived), and breed Y which only produces four piglets per litter but all of which survive. Imagine these two traits are controlled as follows:

Number in litter   A (dominant) = high (16)
                              a (recessive) = low (4)
Percentage surviving B (dominant) = high (100)
                              b (recessive) = low (25)

If pure breeding individuals from breed X (AAbb) and breed Y (aaBB) are crossed the resulting offspring should have the genotype (AaBb). This will result in the offspring producing large litters with a high survival rate. Therefore, cross-bred offspring can be expected to perform better in selected traits than their parents. The amount by which performance in cross-breeds exceeds the parental average is called **heterosis**. Figure 13 illustrates this and table 3.4 shows some examples of heterosis in common farm animals.

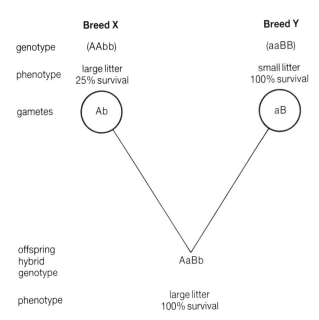

Figure 12 *An example of the inheritance of hybrid vigour*

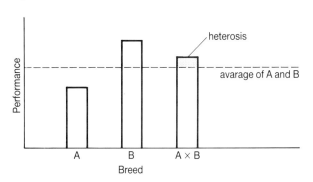

Figure 13 *Cross bred animals tend to perform better in certain traits than their parents. Performance above parental average is called* heterosis. *A graph showing how cross bred animals perform better than their parents*

Table 3.4   *Example of heterosis in farm species*

SPECIES	TRAIT	HETEROSIS %
Dairy cattle	Milk yield	2–10
	Butterfat %	3–15
Beef cattle	Post-weaning gain	4–10
	Carcass composition	0–5
Sheep	Carcass weight	10
	Fleece weight	10
Pigs	Litter size	5–8
	Growth to slaughter	10

(After Dalton 1985)

3.6   MODERN BREEDING PROGRAMMES

Most flocks or herds are relatively small in size. This makes efficient use of natural breeding methods such as cross-breeding difficult as the numbers of animals available for mating are limited. Some breeders overcome this problem by forming co-operative groups which set up a nucleus of breeding animals selected from the member farms. These animals have been selected according to particular traits they possess and form the main breeding herd or flock serving all the member farms. The best animals produced by the unit are returned to the member farms and the surplus sold. This system is more common in Australia and New Zealand than in Britain.

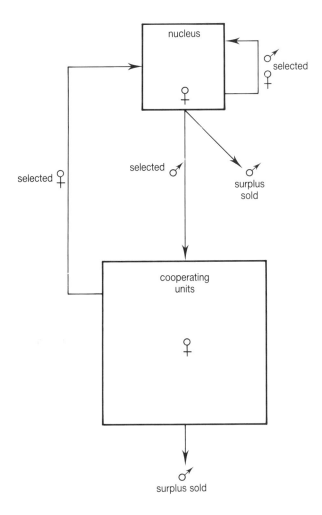

Figure 14 *How co-operative group schemes work*

Many breeders overcome the limitations of herd and flock size by investing in new technologies such as artificial insemination and embryo transfer.

3.7 ARTIFICIAL INSEMINATION

Artificial insemination has been used commercially since the 1940s. It is used with all farm species but has proved particularly advantageous to the dairy industry.

Artificial insemination involves the collection of semen from male animals which have been selected for desirable characteristics. This is usually diluted and stored by freezing. It is then used to inseminate female animals who may also have been selected for the traits they possess. As male animals produce much more sperm per ejaculate than is needed to fertilise the female's ova successfully, many females can be inseminated with the product of one ejaculation. (Artificial insemination is discussed in more detail in Chapter 4.)

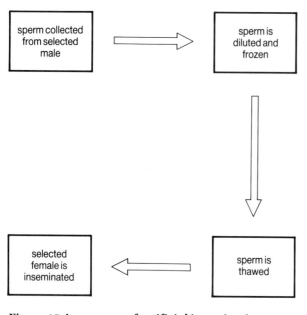

Figure 15 A summary of artificial insemination

3.8 EMBRYO TRANSFER

Embryo transfer has been developed and refined during the last twenty years. The technique involves the extraction of embryos from the reproductive tract of the donor animal and their implantation into the reproductive tract of another animal (the recipient). This enables breeders to select their "best" females and use their ova to produce embryos which will carry the selected genes. Implantation of the embryo into recipients (or non-selected animals) means that pregnancy does not follow in the selected female and so she is free to ovulate again and produce more embryos. Embryo transfer is outlined in figure 16.

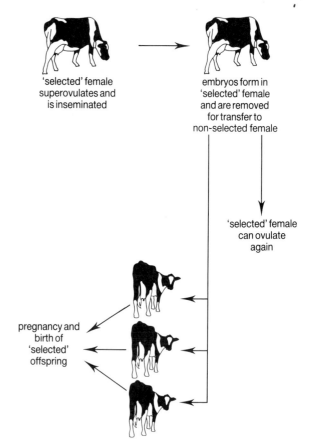

Figure 16 An outline of the process of embryo transfer

3.8.1 Producing embryos for embryo transfer

To make efficient use of the process of embryo transfer, a good supply of selected embryos are required. As the selected donor animal is not going to carry the developing embryos through pregnancy, it is possible to produce many more embryos inside her uterus than normal. To enable the increased production of embryos, donor animals are injected with hormones to make them superovulate. The donor is then mated or inseminated artificially and the embryos are allowed to form.

Embryos can be recovered from the donor animal by surgical or non-surgical means. The age at which the embryo is recovered will determine not only the most suitable method of recovery but also the conception rate in the recipient animals. If embryos are to be recovered soon after fertilisation, surgical methods are required as the embryo will not enter the uterus for several days after it has formed. Donor animals obviously cannot withstand repeated surgery and conception rate in recipients is low for very young embryos and so most embryos are extracted at 6–8 days of age (blastocyst stage) by non-surgical means. Figure 17 shows how conception rate is affected by the age of the embryo in cattle.

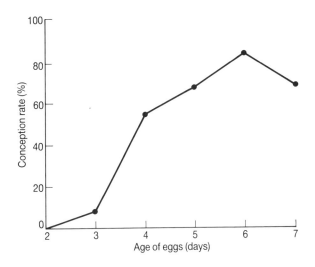

Figure 17 The effect of age of embryo on conception rate after embryo transfer in cattle

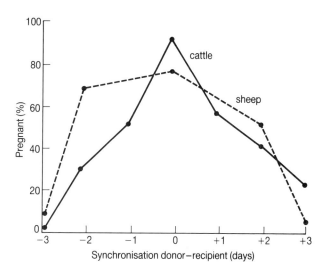

Figure 19 The effects of not synchronsing oestrous cycles in donor and recipient animals, before embryos are transferred

Non-surgical methods of embryo recovery involve the insertion of a catheter, under local anaesthetic, through the vagina and into the uterus. Fluid is then flushed through the catheter, the uterus is washed out and the embryos are collected. In cattle it is possible to obtain up to six batches of 4–6 embryos from each cow, each year, if non-surgical methods are used. This process is shown in figure 18. After microscopic examination, embryos are either transferred immediately or stored by freezing in liquid nitrogen.

3.8.2 Transferring embryos to the recipient

As with collection, embryos are transferred into recipient animals by either surgical or non-surgical methods. Surgical transfer has a higher success rate than non-surgical (55–60% compared with 40–55%) and involves the introduction of the embryo into the uterus by fine Pasteur pipette, through a puncture hole made with a needle. This can be done under local anaesthetic. Non-surgical transfer involves the use of a catheter which is introduced into the uterus through the vagina.

To ensure successful transplantation and development, the recipient animals must be at the same stage of the oestrous cycle as the donor. This is essential if the embryo is to be provided with the uterine secretions required for its development. This can be done by the use of hormones. The graph in figure 19 shows the effects on pregnancy rate of not synchronising oestrous cycles in donor and recipient animals.

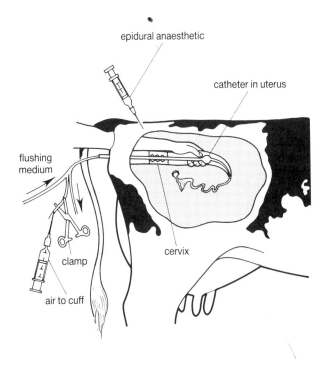

Figure 18 Recovery of embryos by non-surgical methods

Superovulation of donor

Artificial insemination
(5 days after initiating superovulation)

Non-surgical recovery of embryos
(6 to 8 days after artificial insemination)

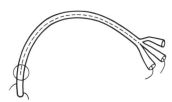

Foley catheter for recovery of embryos

Isolation and classification of embryos

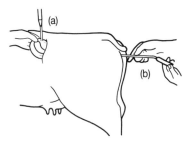

Transfer of embryos to recipients
(a) surgically or (b) non-surgically

Figure 20 A summary of the process of embryo transfer

3.8.3 Advantages and applications of embryo transfer in farm species

The process of embryo transfer obviously allows the desired traits of selected animals to be introduced into successive generations fairly rapidly. Since the development of artificial insemination it has been fairly easy to introduce selected traits from males. With embryo transfer the selection of traits from the female, for example high milk yield, has become more feasible. Examination of the embryos extracted from the donor may allow sexing to take place so that, for example, female embryos can be selected in dairy breeds and male embryos in beef breeds.

It is also possible to use embryo transfer in the production of **clones**. Clones are animals which are genetically identical. It is possible to split the embryo into two inside the zona pellucida which covers it. Half of the embryo is then transplanted into the empty zona pellucida of an unfertilised ova. The two resulting embryos can then be transplanted into separate cows and the calves that they produce will be exact copies (clones). This process is shown in figure 21.

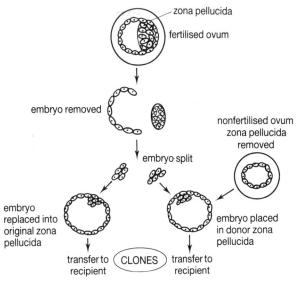

Figure 21 The production of clones

3.9 GENE TRANSFER

The ability to transfer genes from one species to another has only been developed in the last decade. Most of the research into gene transfer has involved mice, but potentially this technique could be used to improve domestic farm animals. When a gene is transferred it will be incorporated into the receiving animal's genotype and so the trait that it controls will show in the receiving animal. So called **transgenic** animals have been produced in several species including sheep and pigs. Human genes have recently been transferred into pigs in an attempt to produce organs which are suitable for use in human transplant surgery.

3.9.1 The process of gene transfer

Gene transfer involves the injection of a gene from one species into another. The most popular way of doing this is to isolate the desired gene and inject it into the nuclei of a fertilised ova using a micropipette. The transferred gene becomes incorporated into the genetic material of the fertilised ova as it develops into an embryo. The resulting animal may be transgenic and carry the desired gene, although the process is not 100% successful.

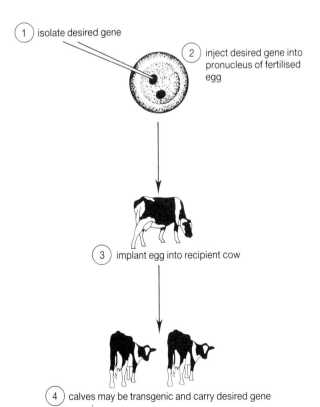

Figure 22 Genes can be transferred by injecting the desired gene into the nuclei of the ova

Alternative methods of gene transfer use viruses. Viruses infect cells by inserting their own genes into the cell's DNA and so taking over the control of the cell. It is possible to introduce the gene to be transferred into the virus so that it is incorporated into the cell by the virus. This is usually done when the embryo is at the 4–8 cell stage. There is a possibility that not all cells are infected and the resulting animals will then control cells which are not genetically identical. These animals are called **chimeras**, and will be a mix of the donor and recipient species.

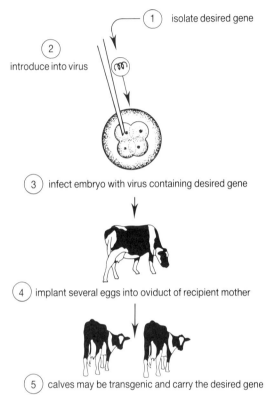

Figure 23 Using viruses in gene transfer

3.9.2 Selecting genes to be transferred

As many characteristics are controlled by a group of genes, the first stage of gene transfer is to select a single gene which controls a desirable trait. This is difficult as only a few single genes, which have useful effects, have been identified. Some possible examples include the gene controlling fertility in Booroola sheep, the halothane gene which controls lean meat content in pigs, and the gene controlling double muscling in some cattle breeds. However, these genes have not yet been isolated and so cannot be used in gene transfer.

Some research has been done into improving milk composition in mice by introducing a gene controlling the whey protein, beta-lactoglobulin (BLG) in sheep milk, into mice. Transgenic mice obtained produced milk with high levels of BLG.

QUESTIONS

1 Suckler herds of cows are often kept for beef production in hilly areas which are unsuitable for other types of agriculture. Calves are kept with and fed by their mothers until they are weaned. The cattle are then sold as store cattle and fattened for market.

The table below shows the heritability of several traits in beef cattle.
a) Which of the traits listed above would you select as desirable if you were:
   (i) The owner of the suckler herd?
   (ii) The breeder who fattens the cattle?
   (iii) The butcher?
b) Which of the traits listed above are most likely to be successfully improved by a breeding programme?
c) The weight of beef cattle at weaning has a low heritability. What environmental factors may influence weaning weight?

TRAIT	HERITABILITY %
Numbers born	5
Birth weight	40
Weight at weaning	25
Post-weaning gain (pasture)	30
Post-weaning gain (store)	50
% lean meat	40
Tenderness of meat	60
Thickness of fat covering	40

2 Improved methods of farming are necessary to provide for the ever-increasing population. For example, the average number of eggs laid per hen in the UK in 1920 was 180 but by 1970 it was 260. This increase has been achieved by breeding programmes involving artificial selection and many generations of inbreeding.
a) What do you understand by the following terms?
   (i) inbreeding,
   (ii) artificial selection.
b) In what way does artificial selection differ from natural selection?
c) Explain two disadvantages of inbreeding.
d) (i) Describe briefly how you would selectively breed a named animal which you have studied. Give the full practical details that would be needed to conduct a breeding programme.

(ii) Explain how you would establish a pure breeding line for a given character.

UCLES 1992

3 To increase the wool density of his flock of hill sheep a farmer considers buying a new ram for breeding purposes.
a) State two pieces of information the farmer needs to know about a ram before buying it to use in his breeding programme.
b) Suggest why starting a breeding programme might be uneconomic for the farmer.

4 The table below shows the performance of Hereford, Angus and cross bred Hereford/Angus cattle on several traits:

BREED	BIRTH WEIGHT (kg)	PRE-WEANING MORTALITY (%)	WEANING WEIGHT (kg)	550 DAY WEIGHT (kg)	% THAT HAVE CALVES
Hereford	34.8	8.6	214.5	305.9	78.2
Angus	31.0	8.6	225.5	305.9	82.3
Crossbred	33.8	3.8	230.5	325.0	93.0

a) For each trait, draw a bar chart to show the performance level of each breed.
b) For each trait, calculate the average performance for the pure bred parents. Draw a horizontal line onto your bar chart to show this level.
c) Calculate the heterosis of each trait.
d) Explain why cross bred animals often show increased performance in particular traits.

5 a) Outline the process of embryo transfer in cattle and explain why the use of this technique has led to great improvement of animals in the dairy industry.
b) The table below shows the effects of not synchronising the oestrous cycle of donor and recipient cows during embryo transfer. Plot this data onto a graph.

VARIATION FROM EXACT SYNCHRONISATION (days)	% OF RECIPIENT COWS FOUND TO BE PREGNANT
0	90
±1	55
±2	35

c) Explain why it is important to synchronise oestrous cycles in donor and recipient cows and state how this is done.

6 a) Describe the techniques used in embryo transplantation.
b) Discuss the implications of the use of such techniques in humans.

UCLES 1992

4 CATTLE PRODUCTION 1: REPRODUCTION

LEARNING OUTCOMES
After studying this chapter you should be able to:
- describe the reproductive system of the bull and discuss how it is adapted to produce and release sperm,
- describe the reproductive system of the cow and discuss how it is adapted to produce ova,
- outline the hormonal control of the oestrous cycle in the cow and relate this to time of insemination,
- outline the process of artificial insemination and compare the advantages and disadvantages of this technique with that of natural insemination,
- describe what happens during pregnancy and birth in cattle.

4.1 DOMESTIC CATTLE

Cattle have been domesticated farm animals for many thousands of years. The first record of their domestication was 6000 BC in Anatolia, Turkey. Cattle are reared for the production of milk, leather and the meats beef and veal. In some countries they are still used for haulage and farm work.

As with other mammals both the male and female reproductive systems do not become fully functional until the animal reaches puberty. This occurs at between 8–18 months, depending on the breed and is often dependent on the mass of the animal.

4.2 THE BULL

4.2.1 The Male Reproductive System

The male reproductive system consists of the **testes**, the **epididymis**, the **vas deferens** and the **penis** as shown in figure 24.

prostate gland

seminal vesicles

Cowper's gland

retractor muscle

sigmoid flexure

vas deferens

penis

sheath

testicles

scrotum

Figure 24 The reproductive organs of the bull

The testes are located within the scrota which hang outside the body. The testes consist of seminiferous tubules – a series of convoluted tubes which are surrounded by interstitial cells. Sperm produced by the testes (see later) are stored in the epididymis where they mature.

The vas deferens transports mature sperm from the epididymis to the penis. Whilst in the vas deferens, chemicals from the accessory sex glands are added to the sperm to produce semen. The accessory sex glands are the **Cowper's gland**, the **seminal vesicles** and the **prostate gland.**

4.2.2 The male gametes – spermatozoa

The spermatozoa or sperm cell is the male gamete in all vertebrate animals. The size and shape of the sperm will depend upon the species from which it is obtained. The general structure and function of the sperm is, however, very similar in all species. Figure 25 compares sperm from a bull and a cockerel and figure 26 shows the general structure of a sperm.

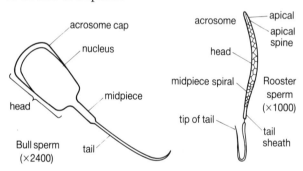

Figure 25 Sperm from a bull (left) and a cockerel (right)

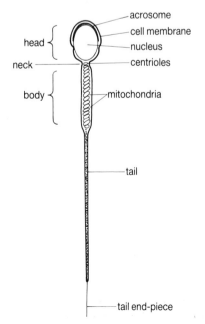

Figure 26 Generalised sperm structure

The sperm is made up of a number of sections. The "head" contains the nucleus and hence the chromosomes, which will determine the characteristics of the embryo it produces. This is discussed in Chapter 2. The nucleus is surrounded by the **acrosome**. This is a membrane containing enzymes which help the sperm to enter the ova during fertilisation.

The "mid" section of the sperm is packed with mitochondria. Mitochondria produce energy in the form of adenosine triphosphate (ATP) which enables the sperm to move. The "tail" or flagellum uses the energy produced and undulates, propelling the sperm at a rate of 1–4 mm a minute.

4.2.3 Spermatogenesis

Sperm production or spermatogenesis occurs in the testis. In most vertebrates, these hang outside the body of the male in the scrota. This helps keep them cool as spermatogenesis is inhibited by high temperatures. The scrota maintain the testes at a temperature 2–3 times cooler than normal body temperature so spermatogenesis occurs efficiently.

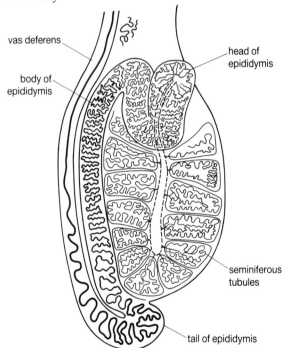

Figure 27 The testes structure

Each testis is made up of the seminiferous tubules. These are highly coiled tubes held together by connective tissue called interstitial tissue. The walls of the seminiferous tubules are extremely convoluted and made up of the **primordial germ cells** and **Sertoli cells**.

The primordial germ cells divide by mitosis to produce identical spermatogonia cells. These grow into primary spermatocytes. The cells undergo meiosis during which they divide twice.

During the first division, the chromosomes within the cell nucleus are duplicated and two cells are formed, each containing a complete set of chromosomes. These cells are called secondary **spermatocytes**. During the second division the chromosomes are not duplicated and so the two secondary spermatocytes produce four **spermatids**, each containing half the number of chromosomes found in the nucleus of the original germ cell. They are said to be **haploid**.

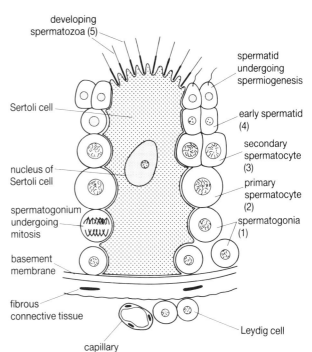

Figure 29 Sertoli cells

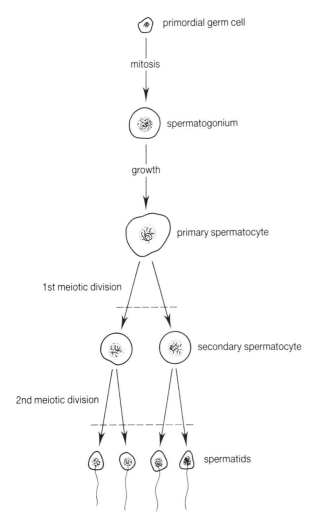

Figure 28 Spermatogenesis

4.2.4 The development of the sperm

The large **Sertoli** cells which are embedded in the wall of the seminiferous tubules provide nutrients for the spermatids so that they are able to mature into spermatozoa. The unspecialised spermatids are attached to the side of the Sertoli cells and eventually differentiate to show characteristic head, mid piece and tail.

The tails of the spermatozoa hang into the lumen of the tubule and the heads are embedded into the Sertoli cells. Once the spermatozoa has reached maturity it is released into the lumen.

4.2.5 The storage of sperm

Sperm produced in the seminiferous tubules are stored in the epididymis. The epididymis is a mass of ducts found on the outside of the testis, which leads to the vas deferens. Sperm can be stored within the epididymis for several weeks during which time they are nourished by fluids secreted from the walls of the ducts. If storage is prolonged or overheating occurs, the sperm will begin to die.

4.2.6 The production of semen

During ejaculation, sperm is propelled from the seminiferous tubules and epididymis, through the vas deferens and urethra and out through the erect penis. As sperm travels through the vas deferens it is mixed with fluids from the seminal vesicles, the prostate and Cowper's glands to produce a milky fluid called **semen**. These fluids contain nutrient for the sperm and stimulate them to swim. The volume of semen produced and the numbers of sperm it contains varies with species as shown in table 4.1.

The bull produces between 2–10 ml of semen per ejaculation. Each millilitre contains up to 2000 million sperm, far more than is needed to successfully fertilise an ovum!

Table 4.1 *Volume of semen and numbers of sperm per ejaculate in different species*

SPECIES	AVERAGE VOLUME (ml)	SPERM (ml×10⁶)
Cattle	2–10	300–2000
Sheep	0.7–2	2000–5000
Goat	0.6–1	2000–3500
Horse	30–300	30–800
Pig	150–500	225–300
Chicken	0.2–1.5	0.5–60
Turkey	0.2–0.8	0.7

(From *Reproductive Physiology in Mammals and Birds.* Nalbandov, AV 1976.)

4.2.7 Hormonal control of spermatogenesis

Spermatogenesis is stimulated by the production of the hormone FHS (follicle stimulating hormone). This is produced by the anterior pituitary gland at the base of the brain in both males and females. The anterior pituitary gland also produces a second hormone, interstitial cell stimulating hormone (ICSH). This stimulates the interstitial cells in the testes to produce testosterone – the male sex hormone. Testosterone is responsible for regulating the development of the epididymis, the vas deferens and the penis, and for the development of the secondary sex characteristics.

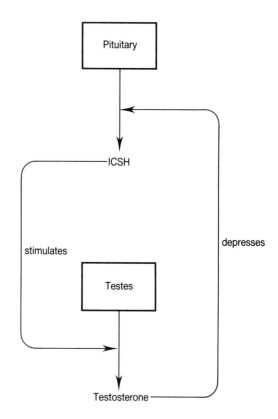

Figure 30 The mechanisms of hormonal control of testosterone

4.3 THE COW

4.3.1 The reproductive system of the cow

The female reproductive system is far more complicated than the male. A outline is shown in figure 31.

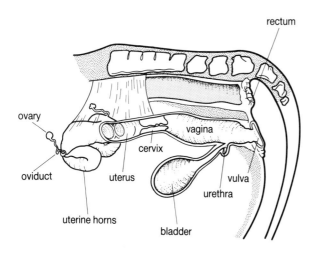

Figure 31 The reproductive organs of the cow

The ova develop within the ovaries. These weigh about 10–20 g and are oval in shape. The ova exist as immature primary follicles which develop into ovarian follicles as they mature.

When ovulation occurs the ovum travels down the Fallopian tube to the uterus. Fertilisation will take place during this journey and the resulting embryo will implant in the uterus. The uterus is shaped like a ram's horns with a central chamber between them (bipartite). The Fallopian tubes lead into the horns and this is where the embryo develops.

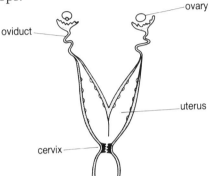

Figure 32 The bipartite uterus in the cow

The cervix is a sphincter-like structure which separates the vagina from the uterus. The main function of the cervix is to seal off the uterus so that pregnancy can be maintained and the uterus remains free from infection.

4.3.2 The female gametes – the ova

The eggs or **ova** are the female gamete cells in all vertebrate animals. They are produced in the **ovaries** and exist in a dormant state from birth. Ovum development is controlled by hormones which are produced by the ovaries and the brain. These also control ovum release or ovulation. Ovulation must occur whilst the male's sperm is inside the body of the female and is still viable if fertilisation is to be successful. It is, therefore, an advantage for the breeder to be able to control and monitor ovulation.

4.3.3 Oogenesis

Ovum production or oogenesis occurs in the ovaries. The ovary is surrounded by an epithelium containing **primordial germ cells**. These germ cells divide by mitosis to produce identical **oogonia**. The oogonia become surrounded by **follicle cells** which also develop from the epithelium of the ovary. The primary follicles contain the **primary oocytes** which are formed and are stored within the ovary until their further development and release is stimulated by hormones.

At puberty, primary oocytes start to develop inside their follicles. In species like cattle which generally only produce one offspring at a time, oocytes develop one at a time. In species, such as pigs which produce several young at once, a number of primary oocytes develop at any one time. In cattle, puberty is reached at about 10.5 months (range 8–17 months). This is fairly late when compared with other farm species as shown in table 4.2.

Table 4.2   Puberty in farm animals

SPECIES	AGE AT PUBERTY (RANGE IN MONTHS)	AGE AT PUBERTY (MEAN) MONTHS
Cow	8.0–17.0	10.5
Sheep	4.5–15.0	7.5
Pig	5.0–8.0	7.0

During the final stages of oogenesis, the follicle cells divide and mature to form a fluid-filled **ovarian follicle** inside which the ova will develop. The primary oocyte divides by meiosis. During the first division of meiosis, the chromosomes within the oocyte nucleus are duplicated and two cells are formed, each containing a complete set of chromosomes. One cell forms the **secondary oocyte** and the second cell forms a small cell called a **polar body**. The first division of meiosis occurs just before the oocyte is released from the ovary (ovulation).

During the second division of meiosis, the chromosomes do not double and so the resulting cells contain only half the number of chromosomes found in the nuclei of the secondary oocyte. This division occurs as the secondary oocyte arrives in the oviduct. The secondary oocyte divides to form the mature **ovum** and a polar body. This is summarised in figure 33.

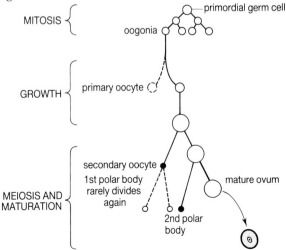

Figure 33 The formation of the ovum

4.3.4 The hormonal control of oogenesis and ovulation

Like spermatogenesis, oogenesis is stimulated by the production of the hormone FSH (follicle stimulating hormone). Unlike spermatogenesis, which occurs constantly, oogenesis is a more controlled process. There are two reasons for this:

(i)   Many animals, especially those living in the wild, reproduce during a particular season. For example, sheep only ovulate in the autumn so that lambs are born in the spring when the weather is warmer and there is more food available. The shorter day-length during the autumn stimulates the production of FSH.

(ii) In mammals, young develop inside the mother's uterus. The size of the developing embryos and the uterus will, therefore, limit the number of young which can be carried at any one time. If oogenesis was a constant process the number of ova available for fertilisation might greatly exceed the space available for the embryos to develop.

In cattle, as in the majority of vertebrates, ova are produced at regular intervals in a cyclic manner. Oogenesis and ovulation are controlled by the release and interaction of FSH and other hormones produced by the ovaries. The cycle of events which these hormones produce is called the **oestrous cycle** and will be discussed later in this chapter. Thus it is possible to detect ovulation in many species and so determine the most suitable time for mating to take place.

4.4   THE OESTROUS CYCLE

(Note the correct spelling of oestrous and oestrus: cows exhibit *oestrus* during part of the *oestrous* cycle.)

As with other mammals, puberty in female cattle is marked by the onset of the **oestrous** cycle. This cycle is controlled by the female sex hormones and begins with a period of **heat** or **oestrus** when the animal is reproductively receptive and able to conceive. If she does not conceive during this time, the oestrous cycle will continue. The length of the oestrous cycle in cows is 21 days. Cows are **polyoestrous**, which means that one oestrous cycle leads directly to the next. **Monoestrous** animals, such as the dog, which have a single oestrous cycle two or three times a year, and sheep are seasonally polyoestrous and have a period of reproductive inactivity.

4.4.1 Hormonal control of the oestrous cycle

The oestrous cycle is regulated by hormones released by the pituitary gland at the base of the brain. (See figure 34.)

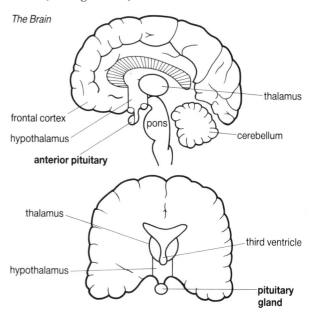

Figure 34 The pituitary gland

Table 4.3   *Length and duration of oestrous cycle, timing of heat and ovulation*

SPECIES	DAYS OF CYCLE	DURATION OF HEAT (hrs)	TIME OF OVULATION	SUGGESTED TIME OF MATING
Cow	21	12–18	12–15 hrs after heat	4–18 hrs before the end of heat
Sow	20–21	48–72	18–40 hrs after the start of heat	24 hrs after the start of heat
Ewe	16–17	24–36	18–26 hrs after the start of heat	12–18 hrs after the start of heat
Goat	19–20	34–39	9–19 hrs after the start of heat	Alternate days during heat
Mare	19–23	90–170	1 day before to 1 day after heat	Alternate days during heat

(From *Reproductive Physiology in Mammals and Birds*, Nalbandov, AV 1976)

Early in this chapter, we discussed the role of follicle stimulating hormone (FSH) from the pituitary gland in the formation of gametes (gametogenesis). FSH also causes the ovaries to secrete another hormone **oestradiol**. At the beginning of each cycle one, or sometimes two, follicles containing a primary oocyte mature within the ovary. It is the maturing follicles that are responsible for the secretion of oestradiol. An increase in levels of oestradiol causes the period of oestrus or heat to begin. The cow is only in oestrus for about 16 hours and will be able to mate with a bull or be inseminated artificially during this period.

Oestradiol has several effects. It stimulates the anterior pituitary gland to stop production of FSH and produce another hormone **luteinizing hormone** (LH). LH brings about ovulation or ovum release. Ovulation occurs 12–15 hours after oestrus in cattle. The ovarian follicle moves to the edge of the ovary and the ovum is released into the oviduct (as shown in figure 35). The remaining follicle develops to form the **corpus luteum**. The corpus luteum secretes the hormone **progesterone** which stimulates the growth of the uterus lining (the endometrium). Progesterone inhibits the release of hormones from the anterior pituitary so that further ovulation cannot take place allowing the body to prepare for pregnancy. Figure 36 shows how hormone levels rise and fall during oestrous.

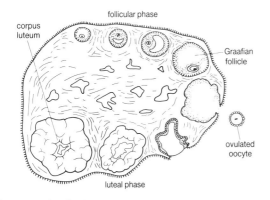

Figure 35 Ovulation

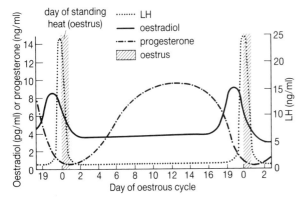

Figure 36 Hormone levels during oestrus

4.5 INSEMINATION

4.5.1 Timing of insemination

The correct timing of insemination is vital to ensure high conception rates. The optimum time is when a cow is in oestrus. This is the time when the cow is willing to stand for mating. If artificial insemination (see later) is used then oestrus detection becomes particularly important and failure to accurately detect it is one of the major causes of poor reproductive performance in cattle.

In cattle, the most obvious sign of oestrus is the acceptance of mounting by other cows. This is called "bulling behaviour". Other signs of oestrus are restlessness, discharge of mucus from the vulva, bellowing, reduced appetite and reduced milk production. Although 95% of oestrous periods can be detected by continuous observation, the average detection rate is only 55–60%. Artificial aids to oestrus detection are available such as paint or phials of coloured dye which can be placed on the tailhead and show when a cow has been mounted. Figure 37 shows bulling behaviour in cattle.

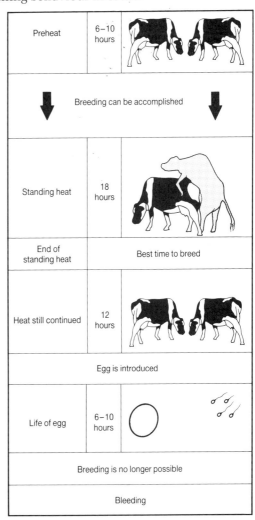

Figure 37 Bulling behaviour in cattle

4.5.2 Natural insemination

Some breeders keep a bull to inseminate their cows. When oestrus is detected the bull will be introduced into the herd. Cows in oestrus will stand and allow the bull to mount them. When sexually excited, blood is pumped into the chambers of the bull's penis. This, and the contraction of the sigmoid flexure muscle causes the penis to become erect so it can enter the cow's vagina as shown in figure 38. The bull ejaculates 2–10 ml of semen, depositing up to 20 000 million sperm inside the female's reproductive tract. Once mating is completed, the penis retracts due to the contraction of the retractor muscles.

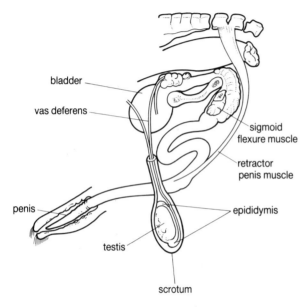

Figure 38 The position of the main muscles controlling the penis

4.5.3 Artificial insemination – advantages and disadvantages

There are many breeds of cattle, each having different characteristics. A farmer owning either a dairy or a beef herd, must be able to ensure that calves born to his cows have the characteristics he requires to enable the herd to produce high yields of milk or beef. Artificial insemination allows the farmer to select particular characteristics which are favourable and ensure that they are passed on from one generation to the next.

Artificial insemination (AI) is more widespread and advanced in cattle (around 70% of the UK dairy herd) than any other species. The main advantage of AI is that it allows more rapid genetic improvement since any farmer can have access to semen from top quality bulls. Such bulls have undergone a progeny test to evaluate the characteristics they will pass on to their offspring.

Other advantages of AI are that it eliminates the dangers and costs associated with keeping bulls, it provides a wider selection of bulls so inbreeding in a herd can be reduced, and it eliminates the transmission of venereal diseases.

The main problem with AI is the need for accurate detection of oestrus to allow insemination at the optimum time — a problem not faced by bulls. Insemination at the wrong stage of the oestrous period, or inseminating cows which are not in oestrus, reduces the conception rate by 2–10%, when compared with natural mating.

4.5.4 The collection of semen

Semen is collected from a bull at an AI centre. Semen is usually collected using an artificial vagina as shown in figure 39. This consists of a rigid container lined with a thin rubber layer. Warm water is placed between these two layers. This causes the rubber layer to stretch and the pressure can be altered to mimic that exerted by the walls of the vagina on the penis during intercourse. Once the semen has been ejaculated it is collected in the glass collection tube.

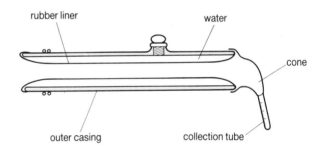

Figure 39 An artificial vagina

Ejaculation may also be stimulated by using an electroejaculator. This is placed into the rectum where it produces a small electric charge which stimulates the reproductive organs causing ejaculation.

The electroejaculator

However, it is the presence of a female in oestrus which normally stimulates mating behaviour and ejaculation in the bull. To mimic this AI centres either use a 'teaser' female which is kept in permanent oestrus by means of hormone injections, or a 'dummy' which is a padded structure impregnated with pungent secretions from other cattle. The dummy can be mounted and the artificial vagina used to collect the semen ejaculated.

4.5.5 Semen dilution and storage

After collection, semen is examined microscopically in order to detect abnormalities in the sperm and to estimate the numbers present. Semen is then diluted because the semen from one ejaculation contains far more sperm than is required to fertilise one ovum. Semen is diluted using a solution which contains nutrients for the sperm, antibiotics to prevent bacterial infection and salts to maintain a desirable pH and osmotic pressure. Bull semen is diluted so that one ejaculation provides material for up to 400 inseminations each containing between $5-15 \times 10^6$ sperm.

The diluted semen is stored by freezing in liquid nitrogen at $-196°C$. It is packed into disposable straws and placed within a storage cylinder as shown in figure 40. Dimethylsulphoxide is added to the semen to help prevent cellular damage during thawing.

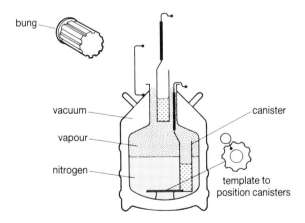

Figure 40 Storage of semen in liquid nitrogen

4.5.6 Artificial insemination

When a cow is to be inseminated, a straw of semen is thawed and placed in a special pipette. A hand is placed into the rectum of the cow in order to position the cervix to receive the pipette. The semen is deposited into the cervix as shown in figure 41.

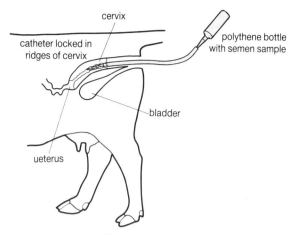

Figure 41 Artificial insemination

> **Now try Investigation 2 Artificial Insemination Techniques in *Animal Science in Action Investigations*.**

4.6 FERTILISATION, PREGNANCY AND BIRTH

4.6.1 Fertilisation

In cattle, sperm can survive for about 30 hours after oestrus, so it is important that insemination occurs towards the end of oestrus if the sperm is to remain viable and fertilise the ovum.

Sperm is deposited at the top of the vagina and propels itself through the cervix and uterus to the oviduct where it will meet the ovum. The head of the sperm makes contact with the ovum, and aided by the acrosome which surrounds the head, enters. The acrosome releases substances which help to breakdown the ovum's surface so that the head can enter. The sperm nucleus fuses with the nucleus of the ovum. As each contains half a set of chromosomes, the resulting zygote contains a complete set of chromosomes carrying traits from each parent.

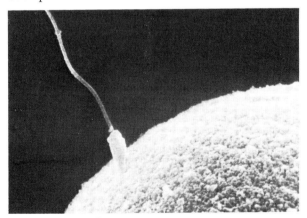

Fertilisation of the ovum

4.6.2 Pregnancy

Pregnancy or gestation lasts for 283 days in cattle. The fertilised ovum floats freely inside the uterus for about 35 days before becoming implanted in the wall of the uterus. Membranes develop around the embryo forming the **placenta** which attaches the developing embryo to the uterus wall. This enables nutrients, oxygen and waste to be exchanged with the mother. It also surrounds the embryo in protective amniotic fluid.

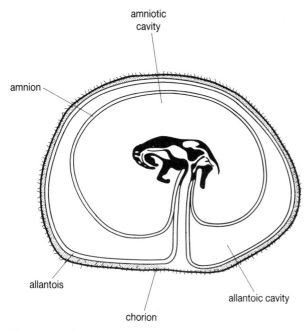

Figure 42 The embryo within the uterus

4.6.3 Pregnancy testing

The primary objective of pregnancy testing is to identify cows that have not conceived and predict when they will return to oestrus. Ideally, this would be done within three weeks of insemination so that only one oestrous cycle is lost in non-pregnant cows. Unfortunately, the earlier pregnancy is diagnosed, the lower the correlation with subsequent production of a calf since embryonic mortality may occur in up to 30% of matings. There are three methods which can be used:

(i) The most widespread method of pregnancy diagnosis is to observe the cow for behavioural signs of oestrus 19–23 days after insemination. If the cow shows no sign of oestrus, she is likely to be pregnant.

(ii) The traditional method used by veterinary surgeons to detect pregnancy is to palpitate the foetus and placental membranes with the hand inserted *per rectum* (see figure 43). This method is more accurate later in pregnancy and is not used before the sixth week of gestation.

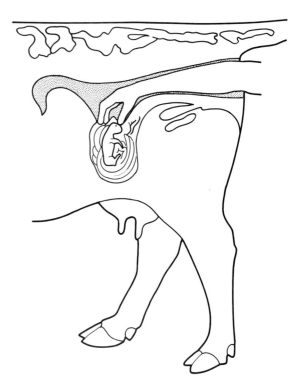

Figure 43 Pregnancy detection by rectal palpitation

(iii) A pregnant cow has higher levels of progesterone in its blood and milk than during the stage of the oestrous cycle when oestrus would occur in a non-pregnant cow. Recently, enzyme kits have been developed so that progesterone levels in milk samples can be measured on the farm. The test is accurate and inexpensive with the added advantage that it can be used in early pregnancy to minimise loss of time in non-pregnant cows.

4.6.4 Calving

Parturition or calving is stimulated by hormones. About six weeks before parturition, the udder will develop. Milk begins to leak from the teats just before the calf is born. Increases in oestradiol production by the placenta stimulates the uterus to produce **prostaglandins**. A hormone called relaxin releases the pelvic muscles so that the calf moves down towards the cervix.

The calf normally appears front feet first, with its head lying between them as shown in figure 44. Its back legs are delivered last. In normal circumstances, the calf will be born within about 30 minutes. If the calf has not been born after two hours, a vet will need to examine the cow to find out if the calf is lying in the wrong position. If this is the case, he will need to reposition the calf inside the cow before it can be born. The placenta is usually expelled 2–6 hrs after birth.

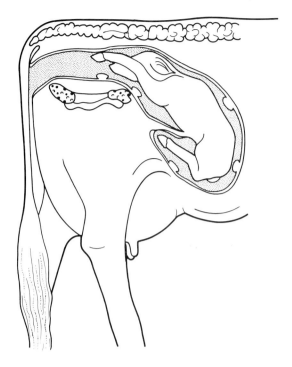

Figure 44 Normal presentation of a calf

The cow will clean her newborn calf, removing parts of the membrane from its face so that it can breathe. The cow will feed her calf within a few hours of birth. The udder produces **colostrum**, a special type of milk which is rich in vitamins, minerals, energy and antibodies. After a few days, the cow will produce milk (lactation). At this point, the calf is usually removed from the cow and fed on formula milk so that the cow's milk can be used for human consumption. Lactation and milk production are discussed in the next chapter.

QUESTIONS

1 a) Sketch the reproductive organs of the bull.
   b) Explain the function of the following:
      (i) The testes,
      (ii) The epididymis,
      (iii) The penis.
   c) Why do the testes in male mammals hang outside the body in the scrota?
   d) Outline the process of spermatogenesis.
2 a) Sketch the reproductive organs of the cow.
   b) Explain the function of the following:
      (i) The ovaries,
      (ii) The oviducts,
      (iii) The uterus.
   c) What causes ovulation to occur in female mammals, including the cow?
   d) The cervix in the cow remains closed most of the time. When must it open?

3 The table below shows the levels of the sex hormones in a cow at different stages of the oestrous cycle:

DAY	PROGESTERONE	OESTROGEN	LUTEINIZING HORMONE
0	0.5	7.0	16.0
2	1.0	4.0	0.5
4	2.0	3.5	0.5
6	5.0	3.5	0.5
8	7.0	3.5	0.5
10	9.0	3.5	0.5
12	9.5	3.5	0.5
14	9.5	3.5	0.5
16	9.0	3.5	0.5
18	4.0	9.0	0.5
20	0.5	7.0	16.0

All hormone levels are given in arbitrary units.
   a) Use the figures shown in the table to draw a graph showing how the levels of these three hormones change throughout the oestrous cycle.
   b) The secretion of these hormones is controlled by another hormone – Follicle Stimulating Hormone (FSH), which is released by the anterior pituitary gland in the brain. FSH causes ova to develop and the ovary to produce oestrogen. Oestrogen stimulates the production of luteinizing hormone (LH) which in turn causes the release of progesterone. Compare the levels of these three hormones on:
      (i) Day 0,
      (ii) Day 2,
      (iii) Day 14,
      (iv) Day 18,
      (v) Day 20.
   c) Oestrus, the period when the cow is receptive to the bull, lasts for about the first 18 hrs of the cycle and ovulation occurs about 9 hrs after the end of oestrus. Which of these three hormones appears to bring about ovulation?
   d) Progesterone is secreted from the corpus luteum. This is formed from the remains of the follicle after ovulation has occurred. Progesterone is responsible for the growth of the uterus lining and its blood supply. What effect will the drop in levels of progesterone at day 15 have?

e) If the cow was to become pregnant, what would you expect to happen to the level of these three hormones? Why?

4 The oestrous cycle is a good illustration of negative feedback. Using it as your example show how negative feedback is essential in maintaining the natural balance within the body of a mammal.

5 a) Why is it important that a breeder knows when the cows in the herd are "on heat"?

b) It is possible to detect oestrus in 95% of cases, however in practice the average detection rate at the farm is only about 60%. What might be the reasons for this?

c) The gestation period in cattle is 282 days so calving occurs approximately nine months after insemination. Most farmers aim to calve about 70% of their cows between August and December and 30% between January and April. At what time of year must insemination occur if calves are to be born:
   (i) In August,
   (ii) In December,
   (iii) In January,
   (iv) In April?.

6 When preparing semen for use in AI it is diluted using a mixture of the following chemicals. Explain why each one is used.
a) Nutrient solution,
b) Buffer solution,
c) Dimethylsulphoxide,
d) Antibiotics.

7 a) List the methods used to detect pregnancy in the cow.

b) What signs would a breeder look for to indicate if a cow was in oestrus? Why is this method inaccurate?

c) Why is rectal palpitation more accurate in the detection of later pregnancy? Why is it dangerous to use this method at stages earlier than six weeks of pregnancy?

BIBLIOGRAPHY

Acker, D. and Cunningham, M. (1991) *Animal Science and Industry* 4th ed. Prentice Hall.

Cohen, J. (1977) *Reproduction.* Butterworth.

Cole, H.H. (1966) *Introduction to Livestock Production.* W.H. Freeman and Co Ltd.

Hunter, R.H.F. (1982) *Reproduction of Farm Animals.* Longman.

Nalbandov, A.V. (1976) *Reproductive Physiology in Mammals and Birds* 3rd ed. W.H. Freeman and Co Ltd.

5 CATTLE PRODUCTION 2: MILK AND MEAT

LEARNING OUTCOMES

After studying this chapter you should be able to:

- distinguish between breeds of cattle commonly used for milk production and those used for meat production,
- describe the process of lactation and explain how the lactation of a cow can be manipulated,
- describe the composition of milk and outline the bonus system which pays farmers a premium for a particular non-fat solids and butterfat content,
- outline the tests available to determine the bacteriological content of the milk and discuss the merits of each,
- distinguish between the techniques commonly used to heat-treat milk and discuss the advantages of each method,
- describe the systems used to produce beef cattle,
- talk about factors which contribute to the overall quality of the meat,
- talk about the production of veal in the UK.

5.1 THE CATTLE INDUSTRY

The cattle industry in the UK is unique because there is a close integration between both the beef and dairy industries. In other countries of the world, specialist breeds have been developed for milk and beef. Male calves born in the dairy herd have very low beef quality and are slaughtered at birth or used to a limited extent for veal production. However, in the UK, over 60% of the one million tonnes of beef produced each year originates from a dairy herd.

5.2 THE DAIRY INDUSTRY

Milk is the main source of food for all newborn mammals. All female farm animals, except poultry, produce milk to feed their young after birth. It is this natural process that has been exploited by humans. There is evidence that cattle were kept for milking as early as 9000 BC, and although there is some market for milk obtained from other species, it is cow's milk that remains the most popular for human consumption. Selective breeding has produced breeds of cow that are very efficient at producing milk.

Table 5.1 Milk-producing ability of farm species

SPECIES	LACTATION (average kg per day)	% OF BODY WEIGHT (per day)
Dairy cow	22	3.2
Beef cow	6.8	1.4
Sow	5.8	3.4
Ewe	1.5	2.2
Goat	3.6	6.0
Mare	11.3–13.6	2.5

The UK dairy industry has an annual output of over £2,000 million and compares favourably with those in other EC countries as shown in table 5.2.

Table 5.2 Milk yield in EC countries

COUNTRY	NUMBER OF COWS	YIELD OF MILK (kg/cow)
Belgium	1153	3866
France	9878	4080
Germany	5762	4630
Irish Republic	1949	3811
Italy	3768	3540
UK	4660	4827

5.2.1 Dairy breeds

Certain breeds of cattle have been selected to produce high milk yields. Those commonly used in the UK are shown below, while table 5.3 shows how the composition of milk varies with breed.

Holsteins (top) were introduced in the 1970s. This breed is large and high yielding and so crossing a Holstein and a Friesian (top middle) produces a very desirable cross breed, suitable for both the beef and dairy industries. Other major breeds in the UK include the Jersey (bottom middle), the Guernsey (bottom) and the Ayrshire which all produce milk with a high butterfat content. There has been a decline in these breeds over the past 20–30 years, probably due to the increased demand for lower fat milks

While table 5.3 shows how the composition of milk varies with breed, 90% of all dairy cattle in England and Wales and 60% in Scotland are Friesian or Friesian crosses.

Table 5.3 The composition of milk from different breeds

BREED	AVERAGE ANNUAL YIELD (kg)	AVERAGE BUTTERFAT (%)
Friesian	6140	3.99
British Canadian Holstein	6998	3.96
Ayrshire	5553	4.01
Jersey	4327	5.50
Guernsey	4490	4.80
Dairy Shorthorn	5429	3.83

(From *Dairy Facts and Figures* 1993)

5.2.2 Trends in the dairy industry

There has been a rapid decline in the number of dairy herds. In the 1950s the number of milk producers reached its peak at 196 000, however, by 1982 this had fallen dramatically to 52 200. The decline has been most rapid in the south and south-east of England, where agricultural land is suitable for other types of farming. However, in areas where dairy farming is still widespread, there has been an increase in the herd size from an average size of 20 in 1960, to 62 in 1982. With increased mechanisation, farmers have been able to increase herd size, and thus profits, and most herds in the UK are now about 100 strong. These trends are shown in figure 45.

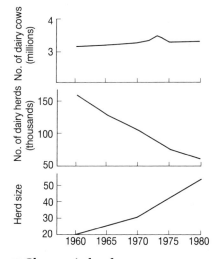

Figure 45 Changes in herd structure 1960–1980

5.2.3 The dairy herd

A cow must produce a calf before lactation (milk production) can begin. Heifers (female cattle before they have calved for the first time) are normally mated so as to produce their first calf when they are between two and three years old. The older they are and the larger they are, the more milk they produce during their first lactation. However, the longer a heifer is kept before she calves, the longer the unproductive phase of her life is and the more expensive the cost of keeping her.

After birth, the calf is left with the cow for one to three days in order to obtain the first milk or colostrum. This is rich in antibodies and provides the calf with immunity to diseases. The calf is then reared on artificial (formula) milk because this can be purchased for £400 per tonne, considerably less than the value of milk which is about £1400 per tonne of dry matter. (Cows are able to lactate and be pregnant at the same time, and so breeders are able to produce a calf a year from each of their cows.) The cow will then produce milk for approximately 305 days. They are then allowed a dry period of about 60 days before they are ready to calve again.

Cows normally remain in the herd for an average of four or five lactations although within any herd cows may range from first lactation to tenth lactation or older. They are culled for a variety of reasons. The commonest is failure to breed, but cows are also culled because of mastitis (a disease of the udder), poor milk yield, lameness, injury, disease or old age.

Cow suckling calf

As cows only remain in the herd for four or five years, heifers must be available to replace 20–25% of the herd each year. Therefore about 60% of the cows in the herd need to be mated to a good quality dairy bull each year to allow for the fact that half of the calves will be male and some will be unsuitable for rearing. The rest of the cows, usually the poorest yielders, are mated with a beef bull and reared for beef production.

5.3 LACTATION

Milk is synthesised in the udder of a cow. This is divided into four separate quarters which are independent of each other. Milk is secreted by **alveoli** which are about 0.2 mm in diameter. These are made up of a layer of epithelial cells which surround a hollow lumen. The epithelial cells extract materials from the circulating blood and use them to produce milk which is secreted into the lumen. The lumens are connected to the **gland cistern** by a series of ducts. This gland can hold up to 1 l of milk at a time. Milk will drain from here to the cavity inside the teat (the **teat cistern**) from where it can be suckled by the calf. Only the small amount of milk stored in the teat cistern is immediately available to the calf. The rest is stored in the ducts and the alveoli and is only drained into the teat cistern when the **let-down** reflex is stimulated.

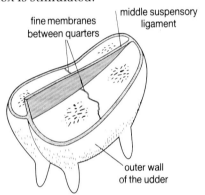

Figure 46 The structure of the cow's udder

Milk let-down is controlled by the hormone **oxytocin**. The stimulation of the calf sucking on the teat (or the milking machine), releases oxytocin from the posterior pituitary gland into the bloodstream. When it reaches the udder it causes the contraction of the alveoli and duct tissue which forces milk into the teat. If the cow is frightened or in pain, let-down is inhibited by the release of adrenalin from the adrenal gland which inhibits the effects of oxytocin.

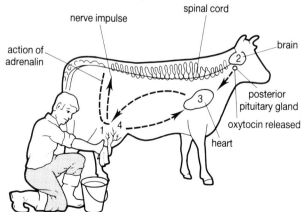

Figure 47 The hormonal control of milk production

5.3.2 The lactation curve

Milk production reaches a peak about six weeks after calving and then declines at a rate of about 2% a week. The rate of decline will depend on many factors including age, nutrition and frequency of milking.

As a cow is able to lactate and be pregnant at the same time the herd is usually managed so that each cow produces one calf per year. This means that the cow has a 60 day drying-off period before the birth of the next calf. The drying-off period is initiated by ceasing to milk the cow so she stops producing milk. Her body is then able to build up reserves for the next lactation once the new calf is born. It is, therefore, desirable for the cow to become pregnant about 83 days after the birth of her last calf.

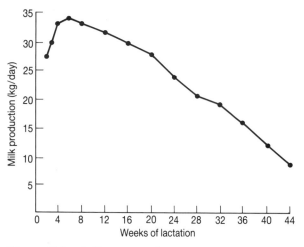

Figure 48 Lactation curve of a cow

5.4 THE COMPOSITION OF MILK

The average milk yield of a cow in the UK is 5000 litres per year, and ranges from 3000–10 000 litres. The determinant of potential yield is the genetic merit of the cow and her ability to reach that potential is dependent upon the quality and quantity of food available. Generally, cows given large quantities of cereals and good quality protein, such as fishmeal or soya-bean meal, produce most milk. Those that are fed mainly on forages, such as silage, produce less milk, but, because the cost of forages is less, such systems can be just as profitable.

Milk is produced in the udder from materials which are extracted from the blood. To provide one pint of milk, 500 pints of blood must be filtered. Table 5.4 shows how the composition of milk varies from the composition of plasma.

Table 5.4   Composition of plasma and milk

CONSTITUENT	% IN PLASMA	% IN MILK
Water	91.00	87.00
Lactose	0.05	5.00
Protein	7.50	3.50
Fat	0.06	3.60
Calcium	0.01	0.10
Phosphorus	0.35	0.10
Sodium	0.35	0.05
Potassium	0.03	0.15

Milk is the single food which most completely meets the needs of the growing animal. Its nutritional value is very similar to its composition and it is easily digested.

5.4.1 Milk composition

Milk is one of the most nutritious foods known to man and every precaution is taken in the progression from cow to consumer to ensure that the quality of milk is of the highest standard. The quality of a sample of milk is assessed in a number of ways. Milk must contain few micro-organisms, be free from antibiotic residues and be of a certain composition.

When a sample of milk reaches the dairy, its composition is tested. Farmers are paid a bonus for milk which has a high "total solids not fat" (SNF) content. This figure tells us something of the protein and carbohydrate levels of the milk. The butterfat level is also checked. A farmer will have his milk checked each month for these two components and the price he is paid for his milk is based on this. Table 5.5 below gives some examples of how the price of milk varies according to the level of SNF and butterfat.

Table 5.5 The compositional quality payment scheme

A. Solids Not Fat

% SNF IN SAMPLES	BASIC PRICE PER LITRE
For each 0.1% above 9.50%	+0.088p
9.40–9.50	+0.616p
9.30–9.40	+0.528p
9.20–9.30	+0.440p
9.10–9.20	+0.352p
9.00–9.10	+0.264p
8.90–9.00	+0.176p
8.80–8.90	+0.088p
8.70–8.80	0.00
8.60–8.70	−0.088p
8.50–8.60	−0.176p
8.40–8.50	−0.264p
8.30–8.40	−0.352p
8.20–8.30	−0.440p
8.10–8.20	−0.528p
8.00–8.10	−0.616p
Less than 8.00	0.704p

B. Butterfat

% BUTTERFAT IN SAMPLE	BASIC MILK PRICE PER LITRE
For each 0.1% above 5%	+0.144p
4.90–5.00	+1.584p
4.80–4.90	+1.440p
4.70–4.80	+1.296p
4.60–4.70	+1.152p
4.50–4.60	+1.008p
4.40–4.50	+0.864p
4.30–4.40	+0.720p
4.20–4.30	+0.576p
4.10–4.20	+0.432p
4.00–4.10	+0.288p
3.90–4.00	+0.144p
3.80–3.90	0.00
3.70–3.80	−0.144p
3.60–3.70	−0.288p
3.50–3.60	−0.432p
3.40–3.50	−0.576p
3.30–3.40	−0.720p
3.20–3.30	−0.864p
3.10–3.20	−1.008p
3.00–3.10	−1.152p
Less than 3.00%	−0.144p

> **Now try Investigation 5 The Constituents of Milk in *Animal Science in Action Investigations*.**

5.4.2 Bacteriological quality

Milk is never completely sterile – milk in the udder contains about 300 micro-organisms per ml. This is a fairly low count but if milk is left untreated they will multiply rapidly. It is important that milk is stored and handled properly during milking and processing so that the numbers of bacteria which enter are as low as possible. Bacteria which grow in milk may cause disease or produce lactic acid from the lactose sugar causing it to go off. A farmer is paid 0.23p more per litre if the milk he delivers to the dairy has less than 20 000 bacteria per ml. If it contains more than 100 000 bacteria per ml the farmer will lose between 1.5p and 10p per litre, depending on the bacterial quality of the milk over the previous six months.

There are many tests performed on milk samples to determine its bacteriological quality.

A. Microscope Counts

(i) **The Breed Smear:** 0.01 ml of milk is pipetted onto a 1 cm^2 area of a microscope slide. The fat is removed and the slide is examined under the microscope. The numbers of organisms per ml can be calculated.

(ii) **Plate Counts:** Milk is diluted using 0.8% saline solution or Ringer's solution and samples are plated onto milk agar. Plates are incubated at 30°C and colonies are counted. This method is inaccurate if milk contains less than 30 organisms per ml or more than 300.

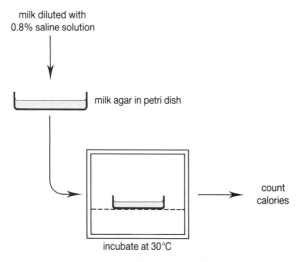

Figure 49 Plate count method of checking milk quality

(iii) **Roll Tube:** Tubes are inoculated with 4.5 ml of agar and 0.5 ml of milk. The tubes are spun so that the agar sets in a fine film coating the tube. After incubation the colonies are counted.

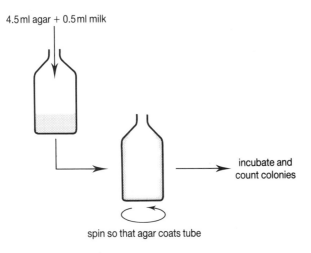

Figure 50 Using the roll tube test to assess milk quality

(iv) **Frost's Test:** A special microscope slide which has an area 20 × 25 mm etched on it, is placed on a hot plate to keep it warm and 0.05 ml of milk and 0.05 ml of agar are mixed on the square. The slide is then incubated at 30°C for 24 hours. The film is then dyed using methylene blue and colonies are counted using a mm^2 grid. It is possible to calculate the numbers of bacteria in the milk sample.

B. Chemical Tests

(i) **Methylene Blue:** Methylene blue is a dye which discolours when it is reduced. During reduction oxygen is removed. If a sample of milk has bacteria living in it they will use up oxygen, causing the dye to be reduced and change colour from blue to white. As the table below shows, different species of bacteria reduce methylene blue at different rates:

Table 5.6   Variation in reduction time of methylene blue with species

SPECIES	REDUCTION TIME (hrs) FOR EQUAL NUMBERS OF BACTERIA
Streptococcus lactis	3.5
Escherichia coli	4.0
Staphylococcus aureus	4.0–6.0
Micrococcus spp.	4.0–6.0
Bacillus spp.	7.0–9.0
Achromobacter spp.	10.0

(ii) **Resazurin Test:** The test works in a similar way to the methylene blue test. Resazurin changes from blue to pink to colourless as the numbers of bacteria in the sample increase. It is a useful test as it can be done quickly when milk is delivered allowing the dairy to accept or reject samples. Resazurin is added to samples and these are placed in a water bath at 37°C for 10 mins. The samples are then compared with a series of standards which can be produced by holding a sample of control milk against coloured glass plates. The milk is then graded on a scale of 1–6.

| 6 – blue |
| 5 – lilac |
| 4 – mauve |
| 3 – mauve/pink |
| 2 – pink/mauve |
| 1 – pink |
| 0 – colourless |

Figure 51 Colour scoring of milk using Resazurin

Samples with a reading of 0–1 are rejected, milk with a reading of 4–6 are accepted. Any samples falling into the range 2–3 will be retested.

C. Antibiotics

Antibiotics are used in the treatment of cows with bacterial infections. Residues of antibiotics can be excreted in the milk for up to five days after treatment has finished. Normally, milk is withheld from sale for the required length of time to allow these residues to clear. Farmers have milk from their farm tested at least once a month. If it is found to contain more than 0.006 International Units of antibiotic per ml, the producer will only receive 1p per litre for the whole of that consignment.

Now try Investigation 6 The Keeping Qualiy of Processed Milk in *Animal Science in Action Investigations*.

5.5 PRESERVATION OF MILK

Milk is a good source of food for bacteria. The bacteria which live in milk can be divided into two groups:
  (i) Pathogens such as *Micrococcus tuberculosis* which cause disease,
  (ii) Saprophytes such as *Streptococcus lactis* which cause the milk to spoil by producing lactic acid.

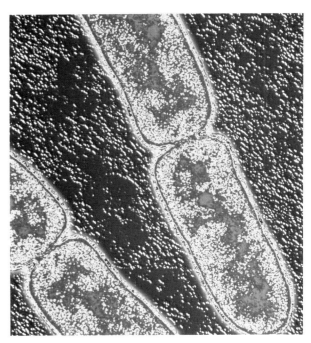

Micrococcus tuberculosis — *a pathogen once common in raw milk*

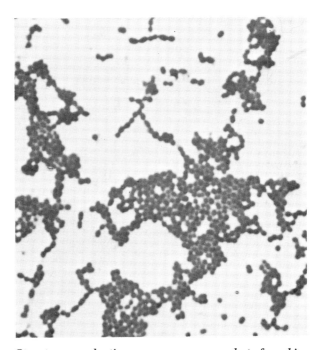

Streptococcus lactis — *a common saprophyte found in milk*

Almost 99% of milk produced for human consumption is heat treated to preserve it. The most common methods of preserving milk is by **pasteurisation** or by **sterilisation**.

5.5.1 Pasteurisation

As most milk is sold for home consumption it does not require a very long shelf life. Pasteurisation is a technique designed to destroy the pathogens in milk so that it is safe to drink.

49

However, many saprophytes remain and so the milk has a fairly low keeping quality.

Milk is heated to a temperature of 71–72°C for 15 seconds. This kills 99% of bacteria but none of the spores which some saprophytic bacteria produce. These spores allow the bacteria to remain dormant and grow when conditions are suitable.

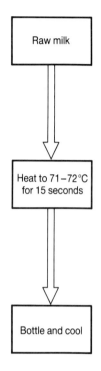

Figure 52 The pasteurisation of milk

5.5.2 UHT – ultra heat treatment

UHT milk is heated to a much higher temperature, 135°C for two seconds, using steam, and then cooled rapidly to 20°C. This kills all spores and bacteria making the milk safe to drink with a good keeping quality. The high temperatures, however, can cause the flavour of the milk to change as it affects the other components of the milk. Before ultra heat treatment can begin, milk is **homogenised**. During this process, the milk is forced through tiny jets to break down the fat particles so that they are able to mix with the watery part of the milk. The milk is then packaged aseptically in cartons (see figure 53) so that no microorganisms can enter. Cartons can be stored unopened for long periods of time without the use of a refrigerator, however, once a carton of UHT is opened, it is susceptible to bacteriological attack.

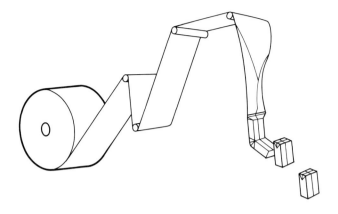

Figure 53 The aseptic packing of UHT milk

5.5.3 Sterilisation

Sterilised milk is produced by placing the raw milk into bottles and heating the whole thing to 130°C for 10–20 minutes. Unopened, this milk has a shelf-life of up to five months as all bacteria are destroyed, however, the flavour and vitamin content are also affected.

> **Now try Investigation 7 Testing the Success of Milk Processing and Investigation 8 Investigating the Activity of Rennet in *Animal Science in Action Investigations*.**

5.6 THE BEEF INDUSTRY

The beef industry in the UK is unique because it is closely integrated with the dairy industry. In most countries specialist breeds have been developed for each of the two industries but in the UK this is not the case. In the UK over 60% of the 1 million tonnes of beef produced each year originates in the dairy herd. Male calves of dairy breeds and calves from low-producing cows which have been mated with a beef bull are reared for beef production. The majority of cows that are culled from the dairy herd can be fattened for beef. Often cows are culled at a young age because of poor performance or reproductive failure, so their meat is good quality. Meat from older cows is normally processed into sausages, pies and other meat products.

The rest of the beef produced in the UK comes from specialised beef breeding or **suckler** herds. These herds consist either of cows that are pure beef breeds or beef cross dairy cows that originate in the dairy herd. Such herds are mainly found in areas of the country that are not suitable for dairy or arable farming.

5.6.1 Beef breeds

As we saw in the last section, 60% of all beef production in the UK comes from cattle originally bred for the dairy industry. The remaining 40% comes from specialised beef breeds or beef breeds crossed with dairy breeds. Some of the most common UK beef breeds are shown below.

1.

2.

3.

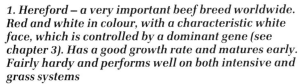

4.

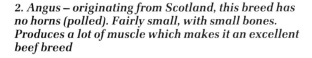

1. Hereford – a very important beef breed worldwide. Red and white in colour, with a characteristic white face, which is controlled by a dominant gene (see chapter 3). Has a good growth rate and matures early. Fairly hardy and performs well on both intensive and grass systems

2. Angus – originating from Scotland, this breed has no horns (polled). Fairly small, with small bones. Produces a lot of muscle which makes it an excellent beef breed

3. South Devon – the largest of the UK breeds, used for both milk and meat production. Produces high milk yields with a high butterfat content (although these are both lower than that obtained in dairy breeds). Can utilise rough grazing

4. Highland – characterised by its long, shaggy coat and long horns. Has a slow growth rate and is late to mature. Very hardy and suitable for poor quality land

Table 5.7   Approximate mature weight of some common UK breeds

BREED	WEIGHT (kg)
Aberdeen Angus	500
Hereford	545
Highland	545
South Devon	660

5.6.2 Beef production

Beef cattle are usually slaughtered between 12 and 24 months old, depending on the system used to keep them (see later). Growth is measured in terms of live weight gain although this may not give an accurate picture of saleable beef production. Growth follows a sigmoidal pattern shown in figure 54, but is affected by nutrition, sex and breed.

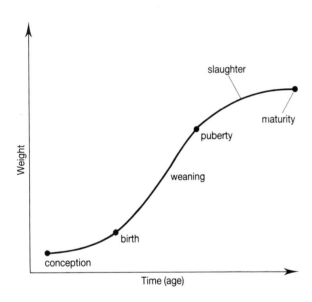

Figure 54 Growth in beef cattle

Producers generally want to produce animals with carcasses that have a high proportion of muscle (meat) and a low proportion of fat. Different types of tissue develop at different rates — nerve tissue develops first, then bone, then muscle and finally fat (see figure 55). Once an animal has reached its maximum muscle production it will convert any extra feed eaten into fat. This is wasteful and so it is important that the producer selects animals for slaughter before too much fat is laid down. A producer can assess the amount of fat cover by handling the five main areas where fat is produced. These are shown in figure 56.

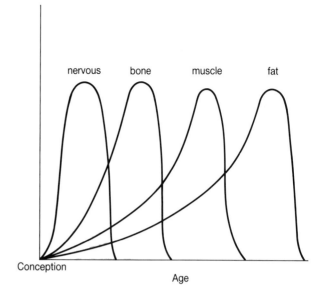

Figure 55 Different types of tissue develop at different rates

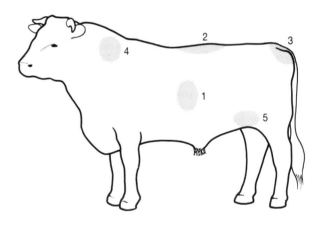

Figure 56 Main areas of fat accumulation

5.6.3 Beef production systems

In Britain there are a number of different systems for beef production which match the type of animal available to the environment and diet offered. A lot of animals are raised in the Western part of Britain where the climate is more suited to grass-growing rather than arable farming. Dedicated beef breeding herds are predominantly found in the hill and upland areas of the country.

Early maturing cattle, which fatten easily, are best suited to systems of production which make maximum use of grass and only obtain moderate growth rates. Such animals take about 24 months to reach slaughter weight at 450–550 kg. If fed on a high-energy diet, they would get fat at too light a weight.

Medium maturing cattle are used in the grass/ cereal system and are slaughtered at 18 months of age, weighing 450–500 kg. They spend their first summer at grass and are fattened in yards on forage and protein and cereal concentrates during the winter.

Late maturing cattle, which require a high-energy diet in order to fatten, are best suited to intensive systems. They are housed indoors throughout their lives and fed on either all concentrates (cereals plus protein supplement) or silage plus concentrates. They reach slaughter weight of 400–500 kg in 12–15 months.

Table 5.8 *Systems of cattle production*

SYSTEM	AGE OF SLAUGHTER (months)	PERCENTAGE OF TOTAL	TYPE OF CATTLE
Cereal	12–15	6	Late maturing breeds
Grass/ cereal	18	49	Medium maturing breeds
Grass	24	45	Early maturing breeds

5.7 MEAT QUALITY

Meat quality is what determines price. Meat quality is monitored by the Meat and Livestock Commission (MLC). Carcasses are classified and grouped into one of six fat classes. These range from 1 (leanest) to 5 (fattest). Most carcasses fall into class 3 and so this has been subdivided into two groups. Table 5.9 shows the composition of carcasses falling into each group.

Table 5.9 *Carcass composition and fat class*

FAT CLASS	1	2	3	4L	4H	5L	5H
% lean	68	67	63	60	59	56	52
% fat	13	14	19	22	25	28	33
% saleable tissue	75	74	63	64	64	61	57
% bone and waste	25	26	37	36	36	39	43

(from Kempster, AJ, Cook, GL, Grantley-Smith, M (1986) Meat Science 17:107–138)

It is the consumer that ultimately determines the price which will be paid for meat, and the farmer must ensure that the quality and composition of the meat produced matches what the consumer desires. Meat quality can be assessed in several ways:

(i) **Taste** Taste cannot be determined when buying meat, however, in general the younger the animal at slaughter, the less taste the meat will have.

(ii) **Tenderness** Tenderness and taste are inversely related so younger animals tend to produce more tender meat. Meat from older animals is less tender because the protein collagen, which is found in muscle tissue, forms cross-links. Meat can be made more tender by 'conditioning'. This involves hanging the meat for a time before eating. Cooking, processing and the use of enzymes e.g. papain, can also make meat more tender.

(iii) **Succulence** Succulence is determined by the amount of fat in the meat. The more fat, the more succulent the meat will be. Beef cattle have fat marbling between muscle fibres, whereas dairy cattle do not.

(iv) **Lean ratio** The lean meat-to-bone ratio is obviously important to the consumer as bone weighs more than meat. A low bone content is therefore desirable.

> **Now try Investigation 3 Identifying the Soluble Amino Acids in Meat and Investigation 4 Meat Processing and Water Holding Capacity in *Animal Science in Action Investigations.***

5.8 THE USE OF HORMONES IN MEAT PRODUCTION

The use of exogenous hormones for beef production was banned by the EC in 1986. This was unfortunate, since anabolic implants increased the efficiency with which animals laid down lean meat with negligible risk to consumers. The use of implants was carefully regulated by legislation. They were put into the ear of the animal, a part which is not used for human consumption. Withdrawal periods during which animals could not be slaughtered were imposed. Most of the hormones used were naturally occurring and consumption of meat from implanted animals posed less threat than consumption of other foods (see table 5.10). Even foods which naturally contain high levels of oestrogen provide little extra exposure when compared with the large quantities naturally produced by humans (see table 5.11).

Table 5.10   Comparative intakes of oestrogen from a variety of food sources

FOOD	WEIGHT OF PORTION (grammes)	OESTROGEN CONTENT (picogrammes)
Untreated beef	500	6 137
Implanted beef	500	11 437
Cow meat	500	7200–538 875
Hen's egg	50–60	1 750 000
Cabbage	100	2 400 000
Peas	100	400 000
Wheat germ	10	200 000
Soya bean oil	10 ml	20 000 000
Milk	500 ml	75 000

Table 5.11   Natural production of oestrogens in humans

	OESTROGEN PRODUCTION $(g \times 10^{-6}/24 \text{ hrs})$
Male child	4
Adult man	136
Female child	54
Non-pregnant woman	540
Pregnant woman	20 000

5.9 BOVINE SPONGIFORM ENCEPHALOPATHY (BSE)

Bovine Spongiform Encephalopathy (BSE), sometimes called "Mad Cow Disease" is a fairly recent discovery and can be an extremely dangerous disease in cattle. The first cases of BSE were noted in the UK in April 1985 and this condition has since been widely reported in the press. In most cases, BSE affects dairy cows of

between 3–6 years of age. Symptoms of the disease include weight loss, a stiffness of the legs, muscle twitching and aggressive behaviour. It is a progressive disease, which results in death in anything between a few weeks to six months after the symptoms first appear. The photo below shows a cow suffering from this condition.

Cow suffering from BSE

It is still unclear what exactly causes BSE. The disease itself is closely related to a disease in sheep called scrapies. This disease is caused by a virus which, in most cases, is transmitted across the placenta from ewe to lamb. It is believed that BSE in cattle is caused by the cattle eating foodstuffs which contain bone meal and meat from infected sheep. The use of sheep waste in cattle feed has been banned since July 1988 in an attempt to reduce the incidence of BSE. There is little or no evidence that BSE can be transmitted from cow to cow or cow to calf.

BSE is a notifiable disease. This means the farmers are, by law, required to inform the authorities of any cattle which may be affected in this way. Cows which are suspected of BSE infection must be slaughtered. The brains are removed for examination and the body is burnt. Cows born before July 1988 are banned from live export as are many calves born from BSE infected cows.

Despite concern over the possible link between BSE in cattle and certain degenerative brain disorders in humans, such as Creutzfeldt-Jakob Disease (CJD), there is little or no evidence of BSE being passed onto another species via the food chain. However, in order to ensure that infected meat does not enter the food chain, food manufacturers are not allowed to use the brain, the spinal cord, spleen, tonsils and intestines of cattle in food products.

5.10 VEAL PRODUCTION

Veal production is a specialised system of calf rearing in which calves are fed milk until they are slaughtered at 14–16 weeks of age. This produces carcasses in the range of 95–120 kg. Veal production is much less important in Britain than in other EC countries and accounts for only 1% of British cattle meat.

Veal calves are usually individually penned, which stops the spread of disease from one calf to another. However, there is increased interest in the systems of group feeding using automatic machines which are thought to be better for the welfare of the animals. There is no truth in the belief that veal calves must be kept in the dark to produce white meat and this practice does not take place in Britain.

QUESTIONS

1 Look at the data in the table below:

Composition of plasma and milk

CONSTITUENT	IN PLASMA %	IN MILK %
Water	91.00	87.00
Lactose	0.05	5.00
Protein	7.50	3.50
Fat	0.06	3.60
Calcium	0.01	0.10
Phosphorus	0.35	0.10
Sodium	0.35	0.05
Potassium	0.03	0.15

a) Illustrate these data visually, using the most appropriate method.
b) What are the major differences between the composition of milk and blood plasma?
c) What do these two sets of figures tell us about the process of filtration which occurs in the cow's udder?
d) The figures shown here are approximate figures. Suggest some factors which might affect the composition of milk.

2 A cow does not produce milk without having a calf and the aim of the farmer is to calve the cow once a year, at a time to suit the system of farming. The graph below shows a cow's lactation over a year.

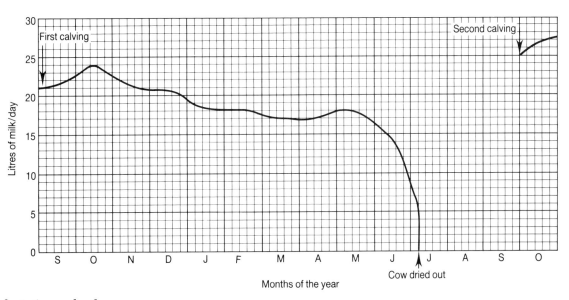

The lactation cycle of a cow

a)
   (i) How much milk was the cow giving per day at peak yield?
   (ii) Calculate the percentage increase in milk yield at the beginning of the first calving.
   (iii) In which month would the cow have to be inseminated in order to produce the second calf in September?
   (iv) State one method by which "drying-off" is achieved.
   (v) Explain how drying-off benefits the cow.

b) Colostrum is produced by the cow for 3–4 days after calving.
   (i) State two ways in which colostrum differs from ordinary milk.
   (ii) State two benefits to the calf of being given colostrum in the first few days of life.

c) State two characteristics which the farmer would consider important in a good dairy cow. For each characteristic, give a reason to support your answer.

London 1992

3 Of the 72 000 calories that a cow consumes each day 18 000 are used up in milk production. A good milking cow produces up to 30 litres of milk a day, whereas a calf only requires about three litres a day.
   a) What proportion of the energy consumed per day is required for milk production?
   b) What proportion of the milk produced by a cow is surplus to its calf's requirements.
   c) How many calories per day does the cow use to produce this extra milk?

4 Use the information found within this chapter to construct a diagram showing how lactation is hormonally controlled.

5 a) List the ways in which micro-organisms may be added to milk after it leaves the udder.
   b) "Clean raw milk" contains micro-organisms from the skin and udder of the cow only. These survive at body temperature. If two samples of milk are compared – one a sample of "clean raw milk" and one a sample of milk which has been contaminated by organisms from soil or milking equipment – explain why the second sample is more likely to "go off" than the first after the milk has been processed and cooled.

6 Write an account of the spoilage of milk by micro-organisms and of methods of preventing such spoilage.

UCLES Modular Science 1992

7 In the UK beef cattle are commonly raised on one of three feeding systems
   • cereals (12–15 months)
   • grass/cereals (18 months)
   • grass (24 months)
   a) Which of the systems above would be classified as
      (i) intensive?
      (ii) extensive?
   b) Why are cattle reared on cereals culled so young?
   c) Why do you think that the cereal/supplement system of feeding is used for only 6% of cattle?
   d) Early maturing cattle kept on grass for most of the year are often moved inside during the worst winter months. During this "store time" they are fed on forage food at a level that just maintains their body. Why is it important that these cattle are not over-fed when in store?
   e) What are the advantages of the grass/cereal system over the other two systems of cattle production?

8 a) Meat quality is assessed by the following factors: taste, tenderness, succulence and lean meat:bone ratio. Rank the quality features listed in order of importance. Explain your ranking.
   b) What other factors may be used to assess the quality of meat?
   c) Which system of beef production would be most suitable for producing "tasty" meat?
   d) Suggest a way in which the enzyme papain may work to make meat more tender.
   e) Which type of cattle would produce the most succulent meat, beef or dairy breeds?

9 Discuss the pros and cons of using hormones in meat production. Assess the effects and potential dangers of treating cattle with hormones. Do you believe that their use should have been banned? Prepare a short speech as part of a group debate, based on your views.

10 Discuss how each of the following contribute to the final quality of beef:
   a) Good husbandry methods,
   b) Good handling prior to slaughter,
   c) Storage and presentation at the point of sale.

UCLES Modular Science 1992

BIBLIOGRAPHY

Acker, D., Cummingham, M. (1991) *Animal Science and Industry* 4th ed. Prentice Hall

Allen, D., Killenny, B. (1984) *Planned Beef Production* 2nd ed. Granada.

Dodsworth, T.L. (1972) *Beef Production.* Pergamon Press.

Leaver, J.D. (1983) *Milk Production.* Longman.

Mepham, B. (1976) *The Secretion of Milk.* (IOB Studies in Biology no. 60) Edward Arnold.

Prestice, T.R., Willis, M.B. (1974) *Intensive Beef Production.* Pergamon Press

Thomas, C., Young, J.W.O. (1982) *Milk from Grass.* ICI Agricultural Division.

POULTRY PRODUCTION 1: MEAT

LEARNING OUTCOMES

After studying this chapter you should be able to:
- list the names of the main breeds of chicken used for egg and meat production,
- outline the processes of broiler production and slaughter,
- list the factors which contribute to meat quality,
- talk about the problem associated with *Salmonella* infections in poultry flocks.

Chicken meat and eggs are both very important parts of the UK diet. British consumers buy about £1300 million worth of eggs a year. That is a consumption rate of about 200 eggs per year or about four a week. The majority of eggs produced come from large flocks of birds (larger than 20 000), however one sixth of all British farmers produce some eggs, usually for local sale.

The consumption of chicken meat in the UK has increased ten fold since the war. Consumers now buy £31.5 billion worth of chicken meat a year. The increase in consumption has been accompanied by a reduction in price (see figure 60) and has been due, in part, to the trends away from eating red meat and to the large variety of chicken products, for example, tandoori chicken pieces, chicken nuggets, etc, which are now available.

6.1 THE POULTRY INDUSTRY

The UK poultry industry is mainly concerned with the production of the domestic chicken for meat and eggs. Other species, for example ducks, geese and turkeys are produced on a smaller scale in certain regions.

Table 6.1 Flock sizes of egg producing chickens in the UK

FLOCK SIZE (hens)	NUMBER OF HOLDINGS (approx.)
0–49	42 000
50–499	6500
500–9999	2400
10 000–19 999	450
20 000+	400
Total	50 750

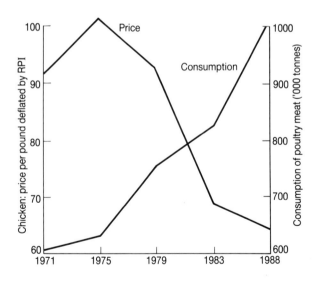

Figure 57 Trends in price and consumption of poultry meat

6.2 BREEDS OF CHICKEN

Traditionally, pure breeds of chicken were used in the poultry industry. However, many hybrid lines are now used. Some breeds are more suitable for egg production than meat production. Breeds of chicken commonly used are shown opposite.

1.

3.

2.

4.

5.

Common breeds of poultry:

1. Single Comb Rhode Island Red: This breed has a yellow skin and red and black feathers. It is a good egg producer, which lays brown-shelled eggs. Many commercial brown egg-laying chickens result from crosses between Rhode Island Reds and other breeds
2. Single Comb White Leghorn: This breed has white feathers and is commonly used in egg production. It produces white-shelled eggs
3. White Plymouth Rock: This breed is yellow-skinned and has white feathers. It is used mainly in meat production
4. Barred Plymouth Rock: This breed has feathers which are grey in appearance due to the black and white bars which run crosswise. It produces brown-shelled eggs

In this section and the next we examine the production of poultry, meat and eggs.

5. New Hampshire: This breed is light red in colour. It has a high rate of egg production, laying light brown eggs, but also produces good quality meat. However, its popularity as a meat producer has declined due to its dark pin feathers which make it difficult to process. The New Hampshire is valuable to the chicken breeding industry as it produces large numbers of eggs which hatch well

6.3 POULTRY MEAT PRODUCTION

6.3.1 Broiler production

Almost all chicken meat is produced in broiler houses. Chickens which are reared for meat production are specially bred and have inherited the ability to grow rapidly and attain market weight quickly. Chickens are brought into the environmentally controlled broiler house as day old chicks. The floor of the house is covered with litter, for example soft wood shavings, which collect droppings. The birds move around the house freely and have constant access to both water and food. At about seven weeks of age the birds have reached a weight of about 1.4 kg/3 lb and are ready for slaughter.

Broiler production

6.3.2 Slaughter and processing

Before birds are slaughtered they are electrically stunned. The bird is hung upside down for a maximum of three minutes and stunned using an electrical current. The throat is cut automatically within 30 seconds of stunning to ensure that the bird does not recover. Birds are allowed to bleed so that much of the blood volume is reduced. This reduces the internal body temperature of the bird and helps reduce the spread of bacteria. The bird is then immersed into a hot water bath at 60°C for 20 seconds to loosen the feathers for plucking.

Birds are plucked automatically by a machine which has banks of rotating rubber "fingers". The feet are then removed before passing onto the next stage, evisceration.

Evisceration involves the removal of most of the organs within the body and the head and neck. This can be done automatically or manually. Evisceration helps to reduce the likelihood of disease spreading from the internal organs to the meat. The bird is now ready for chilling. Birds are chilled rapidly in iced water or air. This is another precaution taken to reduce the spread of bacteria and also reduces toughness (see later).

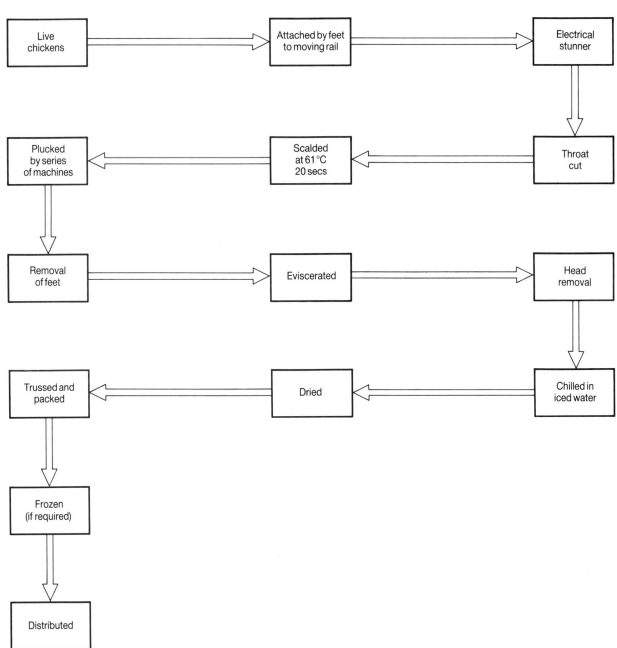

Figure 58 Slaughter and processing of poultry

6.3.3 Meat quality

Meat quality can be assessed in terms of
- tenderness or texture,
- juiciness or dryness,
- bacterial quality.

The **tenderness** or texture of poultry meat is due, in part, to changes which occur in the muscle proteins at slaughter. When a bird is slaughtered **rigor mortis** occurs. During rigor mortis the muscle proteins **actin** and **myosin** combine to form **actinomyosin.** The production of actinomyosin causes the muscles to stiffen and the meat to become tough. In poultry, rigor mortis begins 40 minutes after slaughter. In order to reduce toughness, birds should be chilled to 4–10°C as soon as possible after slaughter.

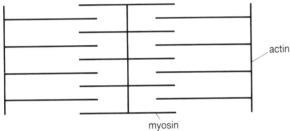

Actin and myosin combine to form actinomyosin

Figure 59 Formation of actinomyosin during rigor mortis

Tenderness is also related to the sex, breed, age and diet of the bird. In general, female birds are more tender than males and younger ones more tender than old ones. Birds which are specially bred for meat production will produce meat which is more tender than egg laying breeds. The ratio of energy to protein in the diet is also important, as high levels of protein in the feed increase tenderness. Most poultry are slaughtered so young that chicken meat is usually tender.

Poultry meat tends to be dry not juicy. To overcome this problem chicken breasts are often injected with polyphosphates after evisceration. This causes them to swell and increases the mass and water-holding capacity of the meat. Most of the polyphosphate is retained when the poultry is cooked and this increases the juicinesss of the meat, although it tends to reduce the flavour. "Organically produced" chickens are not injected with polyphosphates.

Poultry, like all meat, has a short shelf life unless it is processed to preserve it. During slaughter, poultry meat often becomes contaminated with bacteria which cause it to become slimy and rancid. This shelf life of chilled poultry is about seven days when stored at 4°C. This can be extended by freezing or vacuum packing.

6.4 POULTRY AND SALMONELLA

The *Salmonella* group of bacteria are the greatest cause of poultry disease. *Salmonella* infected foods must be strictly regulated as the bacteria can be passed onto humans through both eggs and meat causing food poisoning.

6.4.1 Salmonella infections in poultry

There are two main types of *Salmonella* organism which infect poultry:
- *S. pullorum* which causes a disease known as **bacillary white diarrhoea** (BWD). Birds develop a swollen abdomen, lose their appetite, cheep constantly and have diarrhoea. Mortality can range from 10–20% when the infection is light and husbandry methods are good, to 50%.
- *S. gallinarum* which causes a disease known as **fowl typhoid.** Birds produce greenish-yellow droppings, become thirsty and anaemic. Mortality rates may reach 75%.

Other types of *Salmonella* organisms are also commonly found in poultry, notably *S. enteriditis* which is often present in the guts of birds.

Infected chicks leaving the hatchery will pass salmonella infections on to healthy birds rapidly. The bacteria are found in droppings and can be passed on through contaminated food. Birds carrying common *Salmonella* organisms can be identified by the use of a simple blood test and eliminated. Within the poultry house, hygiene must be strictly controlled. Chicken feed which contains protein from animal sources are often the source of *Salmonella* contamination and so this should be heat treated before use.

6.4.2 Salmonella and food poisoning in humans

Most food poisoning outbreaks are due to *Salmonella* bacteria, usually *S. enteriditis.* They account for half the food poisoning cases reported in Britain every year. *Salmonella* bacteria contaminate foods such as meat, eggs and chicken.

A large number of chickens carry *Salmonella* bacteria in their gut. These are spread to the meat during slaughter and processing. Infestation is limited by rapid chilling and/or freezing. However, if meat is not heated to, and held at, temperatures of at least 56°C when being cooked the bacteria will not be destroyed and food poisoning may result. Salmonella food poisoning is often related to the consumption of frozen chickens which have not been completely thawed before cooking so that the centre of the bird remains below 56°C.

Some types of *Salmonella* bacteria infect eggs whilst they are being produced. The bacteria live in the oviducts and are introduced into the egg before it is laid. Eggs infected with *Salmonella* bacteria require boiling for seven minutes to ensure that bacteria in the yolk are destroyed.

Symptoms of salmonella food poisoning include severe stomach pains, high temperature and diarrhoea. These usually develop a few hours after the infected food has been consumed. Each year a small number of people die from salmonella food poisoning. These are usually from vulnerable groups such as pregnant women, the elderly and the very young.

QUESTIONS

1 The following processes take place during the slaughter and processing of poultry. Arrange them in order and explain the purpose of each stage.
   a) Evisceration.
   b) Bleeding.
   c) Rapid chilling to 4°C.
   d) Electrical stunning.
   e) Scalding in hot water.

2 The table below contains data about the consumption of meat in UK households. The figures in the table show the average amount of poultry, beef, pork and lamb eaten per person per week.

YEAR	POULTRY (g)	BEEF (g)	PORK (g)	LAMB (g)
1975	160	240	85	120
1976	170	210	85	120
1977	175	240	100	110
1978	180	240	100	110
1979	200	240	105	120
1980	180	225	110	125
1981	210	200	105	110
1982	200	200	110	100
1983	205	180	100	110

(From MAFF Data)

a) Plot this data on a graph.
b) Describe the trends which emerge.
c) Suggest what factors may influence the type of meat eaten.

3 The table below shows how certain factors affect the tenderness of chicken meat. The smaller the tenderness measure, the more tender the meat.

BREED	AGE (days)	SEX	TENDERNESS ($\times 10^6$)
White rock	83	male	9.6
	83	female	9.3
	175	male	13.3
	175	female	11.5
Brown Leghorn	105	male	14.7
	105	female	14.6
	203	male	16.5
	203	female	15.5

a) What effect does (i) age and (ii) sex have on meat tenderness?
b) Which of the breeds above is most likely to be bred for meat production, and which for egg production?
c) Some broiler producers add skimmed milk powder to their poultry feed. Why?
d) Meat tenderness is also affected by processing. Explain why meat becomes tough after slaughter and discuss methods by which this can be reduced.

4 The data in the table below shows reported cases of salmonella food poisoning and the number of deaths resulting from salmonella in England and Wales.

YEAR	NUMBER OF REPORTED CASES	NUMBER OF DEATHS
1985	19 241	51
1986	23 806	46
1987	29 626	58

a) Calculate the percentage increase in reported cases of salmonella food poisoning in England and Wales between 1985 and 1987.
b) Suggest some possible reasons for this increase.

ANIMAL SCIENCE IN ACTION

62

c) Calculate the mean percentage death rate from salmonella food poisoning. Most victims are either very old or very young. Why?

d) Prepare a leaflet to educate people about ways of reducing the risk of salmonella food poisoning from eggs and poultry.

5 Describe how you would carry out an investigation to identify the food constituents of a chicken's egg.

London/ULEAC 1992

BIBLIOGRAPHY

Dann, J.E. (1980) *Broiler Rearing.*

Grater, J.M. (1985) *Chicken and Chicken Products – Consumers, Purchasers and Perceptions.* West of Scotland Agricultural College.

Nesheim, M.C., Austic, R.E., Card, L.E. (1979) *Poultry Production* 12th ed. Lea and Febiger.

North, M.O., Bell, D.D. (1990) *Commercial Chicken Production Manual* 4th ed. Van Nostrand Reinhold.

Sainsbury, D. (1980) *Poultry Health and Management* 2nd ed. Granada.

7 POULTRY PRODUCTION 2: EGGS

LEARNING OUTCOMES

After studying this chapter you should be able to:
- describe the reproductive system of the cockerel and explain how it is adapted for the production and release of sperm,
- describe the reproductive system of the hen and explain how it is adapted for the production and laying of eggs,
- outline the formation of each of the layers of the hen's egg,
- describe the composition of the yolk, albumen, shell membranes and the shell and explain how these are altered by storage,
- evaluate and describe methods used to assess the quality of eggs,
- compare methods used for egg production with respect to productivity, animal welfare, and economics,
- outline the process of embryonic development in the chick,
- talk about the conditions which are required for fertilised eggs to hatch successfully.

7.1 THE REPRODUCTIVE SYSTEMS IN THE DOMESTIC CHICKEN

There are many differences between the reproductive systems of the chicken and other fowl, and other domestic vertebrate breeds.

7.1.1 The cockerel

The male reproductive system consists of two testicles which are located high inside the abdomen. In most other vertebrates the internal body temperature would inhibit spermatogenesis.

Sperm are produced rapidly in the seminiferous tubules of the testicles. Mature sperm are stored in the epididymis and the vas deferens which lead from the testicles to the **cloaca**. Sperm have a long pointed head and a long tail. There are no accessory sex glands and so the sperm is mixed only with the seminal fluid from the testicles and fluid produced by the **papillae** – small projections found within the folds of the cloaca. During mating up to 1 cm^3 of semen is ejaculated from the male's cloaca into the female's.

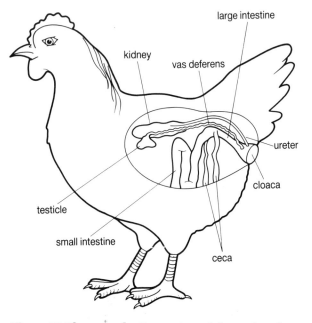

Figure 60 The reproductive system of the cockerel

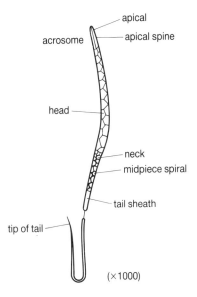

Figure 61 Cockerel sperm

7.1.2 The hen

During embryonic development hens possess two ovaries and oviducts like other vertebrates. However, before hatching, the right ovary and oviduct atrophies (wastes away), leaving only one set. The left ovary contains several thousand ova – only some of which mature into yolks.

The oviduct leads from the ovary. It is a long tube which is normally small in diameter but which is able to increase in size approaching ovulation. The oviduct consists of several sections as shown in figure 62.

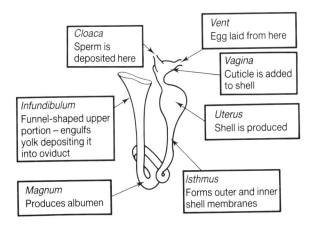

Figure 62 The oviduct of the hen

7.2 EGG FORMATION

The chicken's egg consists of the ova (the gamete cell) surrounded by yolk, albumen, shell membranes, shell and cuticle. These layers are laid down in different areas of the female reproductive tract. It takes about 23–26 hrs for the egg to pass through the oviduct.

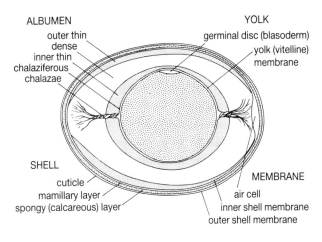

Figure 63 The hen's egg

7.2.1 Yolk formation

The ovary of a chicken can contain as many as 12 000 ova. It takes about ten days for an ovum to develop into a yolk which can be released from the ovary. During this time the yolk grows rapidly to a size of 6–35 mm and is surrounded by the follicular membrane. This attaches the yolk to the ovary and is well supplied with blood vessels so that material for yolk formation can be transported from the bloodstream of the chicken. The yolk is rich in lipids (fat) and protein in the form of low density lipoprotein. It is yellow in colour due to a pigment called xanthophyll which is derived from the food eaten by the chicken. The ovum is located on the surface of the yolk and can be seen as a small white **germinal disc**. If this is fertilised, the growing embryo will derive its nutrients from the yolk.

7.2.2 Ovulation

The developing yolk is suspended from a stalk inside the ovary. This stalk is supplied with blood by an artery which branches to supply blood to the follicular membrane. This blood supply extends over the surface of the yolk except for a narrow band called the **stigma**.

As the yolk matures the ovary produces the hormone progesterone, which stimulates the release of luteinizing hormone from the anterior pituitary gland. This causes the follicular membrane to rupture at the stigma, releasing the yolk into the oviduct. This is ovulation. The yolk is engulfed by the **infundibulum** – the funnel-shaped top of the oviduct. It is then forced into the main oviduct by muscular contraction. The hormonal control of ovulation is summarised in figure 64. Yolk formation and ovulation are also influenced by photoperiodism. Light detected by the eye of the chicken stimulates the production of both follicle stimulating and luteinizing hormones. These stimulate ovarian activity and so stimulate yolk formation. Maximum egg production occurs when 11–13 hrs of light is received. During the winter egg production will fall unless artificial light is provided. This is shown in figure 65.

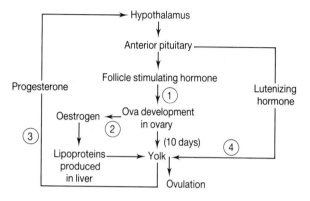

1 Follicle stimulating hormone from the anterior pituitary stimulates yolk in the ovary
2 Ovary produces oestrogen which stimulates the production of lipoprotein for yolk formation in the liver
3 As yolk matures, progesterone level rises causing the hypothalamus to stimulate the production of lutenizing hormone from the anterior pituitary
4 Lutenizing hormone stimulates ovulation

Figure 64 The hormonal control of yolk formation and ovulation

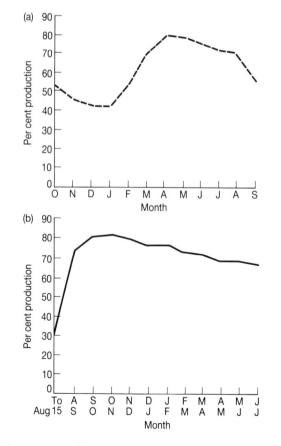

Figure 65 Production patterns from hens when (a) no artificial light is supplied, and (b) when light is supplied

7.2.3 Albumen formation

Albumen or egg white is produced in the section of the oviduct known as the **magnum**. This leads from the infundibulum (see figure 62) and is on average 33 cm in length. The yolk remains in the magnum for about three hours, during which time it is surrounded by albumen. This is all derived from the same type of albumen, however the addition of water and the movement of the egg produces four distinct layers. (Note: albumen is the egg white, albumin a protein found in egg white.)

- **Chalaziferous layer:** this layer of albumin surrounds the yolk. As the egg passes down the oviduct it is twisted. This action causes the formation of two twisted, cord-like structures from this albumin layer. These are called the **chalazae** and extend from the ends of the yolk through the albumen. Their function is to keep the yolk central within the egg after it is laid.
- **Inner thin albumen:** this has a high water content.
- **Dense albumen:** this makes up the largest part of the total albumin. It contains a substance called **musin** which holds it together.
- **Outer thin albumen:** this layer also has a high water content. Outer thin albumen is deposited outside the shell membrane and is the first stage in shell formation (see later). The inner and outer thin albumen are produced by the addition of water to dense albumin.

Table 7.1 Albumen layers in eggs

ALBUMEN TYPE	% OF WHOLE ALBUMEN
Chalaziferous	2.7
Inner thin albumen	17.3
Dense albumen	57.0
Outer thin albumen	23.0

7.2.4 Forming the shell membranes

The shell membranes surround the albumen and are formed in the **isthmus**, the next section of the oviduct. The inner and outer shell membranes are formed in the shape of the final egg and so the contents do not completely fill the membranes at this stage. The outer shell membrane is much thicker than the inner shell membrane as shown in the table. They usually stick together except at the large end of the egg where they are separated by the air cell. This develops after laying and increases in size as the egg ages and the interior dries out. The main function of the shell membranes is to act as a barrier against attack by micro-organisms.

Table 7.2   The thickness of the shell membranes

	THICKNESS (mm)
Outer shell membrane	0.05
Inner shell membrane	0.015

7.2.5 Shell formation

Egg shell is produced in the uterus. This process takes up to 20 hrs. A layer of albumin (the outer thin albumen) is deposited around the shell membrane. This provides a protein and polysaccharide matrix on which the shell can be "built". The shell itself is composed mainly of calcium carbonate with small amounts of magnesium, sodium and potassium. The calcium carbonate required for shell formation is obtained from the diet or from the calcium reserves found in the chicken's bones.

A laying hen producing one egg a day requires 2.5 g of calcium a day for shell production. If this is not obtained from the diet it utilises calcium stored in the bones. One egg requires 10% of the hen's calcium store for eggshell production. A commercial hen may produce up to 300 eggs a year. This puts a great strain on her calcium reserve. If calcium is not replaced the birds may suffer from osteoporosis, a bone wasting disease. Loss of calcium from the bones causes them to become weak and more susceptible to breaking. To help prevent this most flocks are fed supplements containing 3–4% calcium.

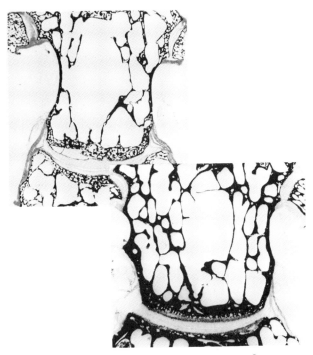

Bones from a bird with osteoporosis (top) and a normal bird (bottom)

7.2.5 The cuticle

The cuticle surrounds the egg shell. It is produced in the uterus and has a high water content to lubricate the egg during laying. After laying the cuticle helps block the pores on the surface of the shell thus reducing water loss and the entry of micro-organisms.

7.2.6 Egg laying

The completed egg now enters the vagina where it is usually only held for a few minutes. Eggs usually travel through the oviduct small end first. The egg is now rotated horizontally through 180° so that it is laid large end first. Egg laying or **oviposition** is stimulated by the hormone oxytocin. This is produced by the posterior pituitary and causes the uterine wall to contract, pushing the egg from the vagina and out through the cloaca.

Chickens will lay eggs on several successive days and then will not lay for a day or two. The successive days of egg production are called a clutch. The size of a clutch will vary with the individual from two to 200 days as will the length of break between clutches. Most commercial chickens produce clutches of between three and eight days.

7.3   EGG COMPOSITION AND QUALITY

Figure 63 shows the structure of a newly laid egg. Hen's eggs weigh about 57 g, are about 7 cm long and have a surface area of 68 cm².

Table 7.3 shows the relative proportions of the three main structures within the egg – the yolk, the albumen and the shell. As the egg is stored the proportions of these components change resulting in a smaller percentage of albumen and a larger yolk.

Table 7.3

COMPONENT	MASS (g)	% OF TOTAL NEW LAID EGG	% OF TOTAL STORED EGG
Yolk	18.4	31.9	35.0
Albumen	32.8	57.1	52.5
Shell	6.3	11.0	11.0

7.3.1 Egg grading

Eggs are graded for sale according to their quality and size.

Eggs can be graded into three classes according to their quality. The class will determine their use.

Table 7.4   Egg grading

COMPONENT	CLASS A	CLASS B	CLASS C
Cuticle	Clean	Dirty or damaged	Dirty or damaged
Shell	Clean and undamaged	Undamaged	Damaged or cracked
Air space	< 6 mm	< 9 mm	> 9 mm
Albumen	Clear, gelatine-like	Clear	Clear, not discoloured
Yolk	Visible as a shadow only when candled	Visible as a shadow only when candled	Distinct on candling
Use	Fresh eggs	2nd quality fresh or processed eggs	Processed

Class A and B eggs are also graded according to their size. It is this parameter that is used when marketing eggs. Eggs which do not qualify for class C are used for industrial processes rather than human consumption.

Table 7.5   Egg sizes

SIZE	MASS (g)
1	> 70
2	65–70
3	60–65
4	55–60
5	50–55
6	45–50
7	< 45

The size of the egg laid by the hen will be determined by a number of factors:
- Genetic influences – some hens exhibit characteristics which contribute to the production of large eggs, for example the production of large yolks in the ovary.
- Hen's age – older hens tend to lay larger eggs.
- Sequence in clutch – eggs produced at the beginning of the clutch tend to be larger than those produced at the end due to the reduced size of both the yolk and albumen.
- Protein content of the feed – a high protein content in the hen's feed will result in the production of larger eggs.
- Climatic conditions – high temperatures tend to reduce egg size.

7.3.2 Factors affecting egg composition

Eggs are made up of water, protein, lipids, vitamins and minerals. The general composition of a newly laid egg is shown in table 7.6.

Table 7.6   The composition of a newly laid egg

COMPONENT	% IN EGG WITH SHELL
Water	65
Protein	12
Fat	11
Carbohydrate	1
Ash	11

The composition of eggs is influenced by the hen's age. Older hens lay eggs within larger yolks but smaller amounts of albumen. The diet of a hen can also affect egg composition. It is possible to alter the mineral and vitamin content of eggs by controlling diet. Other influences on egg composition include the breed of hen and other genetically controlled factors.

7.3.3 The composition and structure of the yolk

The yolk of a newly laid egg contains about 50% water. As the egg is stored this will rise as water migrates from the albumen into the yolk. The remaining constituents of yolk are mainly protein and lipids (fat), with small amounts of carbohydrate in the form of glucose, fat soluble vitamins and minerals (mainly iron and calcium).

Table 7.7   The composition of the yolk

COMPONENT	% OF TOTAL
Water	47.5
Protein	17.4
Lipids	33.0
Carbohydrate	0.2
Minerals and vitamins	1.9

The primary function of the yolk is to provide nutrients for the developing embryo in fertilised eggs. If fertilised, the embryo will develop from the pregerminal disc on the surface of the yolk. It is produced in a series of concentric layers as shown in the photo.

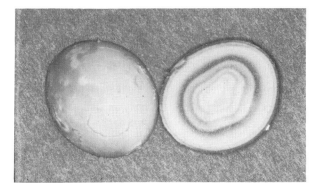

The yolk is laid down in layers as shown by use of a fat-soluble dye

The major component of yolk is **lipoprotein**, a complex of lipids, phospholipids, cholesterol and proteins. These components are believed to be linked together to form large lipoprotein globules as shown in figure 67. The neutral lipids make up the core of the molecule and are surrounded by the phospholipids and protein molecules. These globules are suspended in a clear, yellow fluid along with smaller inorganic granules.

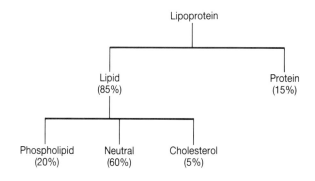

Figure 66 The composition of lipoprotein

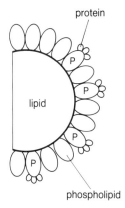

Figure 67 The structure of lipoprotein

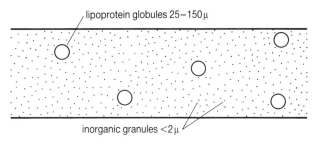

Figure 68 The microstructure of the yolk

The yolk is bound by the **vitelline** membrane. This is composed of an inner layer of non-collagenous protein and an outer protein layer made up of a mixture of proteins – **conalbumin**, **ovadin** and **lysozyme.** Diffusion of materials from the yolk to the white is prevented by an impermeable membrane 0.1 μm thick, which lies between the two layers.

7.3.4 Yolk quality and colour

Yolk quality is determined by measuring the **yolk index.** The yolk index is a measure of the ratio between yolk height and diameter. Both yolk diameter and height will decrease with age, especially at elevated temperatures.

$$\text{YOLK INDEX} = \frac{\text{YOLK HEIGHT}}{\text{YOLK DIAMETER}}$$

Now try Investigation 9 Egg Yolk Quality and Storage in *Animal Science in Action Investigations.*

The colour of the yolk is very important to the consumer, most prefer golden-yellow as opposed to either pale yellow or deep orange yolks. Yolk colour is due to carotenoid pigments (e.g. xanthophylls, lutein and zeaxanthin) within the yolk. The hen is unable to synthesise these pigments and so they are obtained from the feed that the hen eats. Several feedstuffs provide relatively high xanthophyll levels as shown in table 7.8.

Table 7.8 The xanthophyll content of poultry feedstuffs

FEEDSTUFF	XANTHOPHYLL mg/kg
Yellow corn	17
Corn gluten meal (60% protein)	290
Alfalfa meal (20% protein)	280

Yolk colour can be measured either subjectively or objectively.

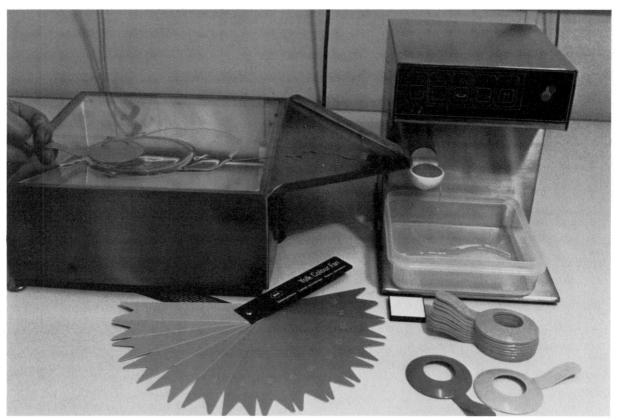

Ashton-Fletcher colour rings, Roche yolk colour fan and the yolk colorimeter, the latest technology in objective measurement of yolk colour

Subjective assessment of yolk colour

The subjective assessment of the colour of egg yolk depends on comparing the colour of egg yolks with prepared standards.

(i) **Heiman-Carver Colour Rotor** (1935) This consists of 24 painted watch glasses which range in colour from pale cream to orange-red. Eggs are placed underneath the wooden disc on which the watch glasses are mounted and the colour of the yolk is compared with each coloured watch glass.

(ii) **The Roche Yolk Colour Fan** (1956) This consists of 15 colour standards which correspond to the range normally found in egg yolks. These standards can be used to compare yolk colour in samples.

(iii) **Ashton-Fletcher Colour Rings** This system consists of 15 concave aluminium rings each of which are coloured with a standard yolk colour ranging from light yellow to deep orange. The rings have a hole punched in the centre so that they can be held up directly over the yolk for colour matching.

(iv) **Bornsten and Bartov Dichromate Solutions** (1966) The colour of yolk extracted in acetone is compared with standard dichromate solutions (1–12 mg/ml water) against a white background. This method has the advantage that it is relatively inexpensive and the colour of dichromate solution remains stable for up to 12 months so standards do not have to be replaced.

These visual methods of assessment are inexpensive, quick and relatively easy to perform. However, there are many problems associated with those subjective assessments which make them inaccurate.

- Egg yolks are translucent – standards are usually opaque.
- Egg yolks are not flat unlike many of the standards.
- Standards may fade with storage.
- Different observers may interpret colours differently.

The speed at which this type of assessment can be performed does, however, make it suitable for quality control.

Now try Investigation 10 Comparing Yolk Colour in Battery and Free-Range Eggs and Investigation 11 Assessing Yolk Colour Preference in *Animal Science in Action Investigations*.

Objective assessment of yolk colour

(i) Using the spectrophotometer to measure pigment concentration.

Yolk pigment can be extracted in acetone. This dissolves the lipid in which the pigment is stored. This solution can then be compared with standard β-carotene solutions in the spectrophotometer, which shines light through the sample and measures the amount of light absorbed by the sample. The reading obtained is the **optical density.** If the spectrophotometer is calibrated using known concentrations of β-carotene it is possible to produce a standard curve from which readings obtained for solutions of extracted egg yolk can be converted into micrograms of β-carotene per gram of yolk. This method is fairly accurate but very time consuming.

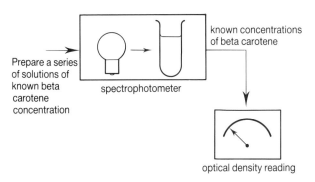

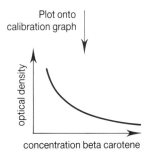

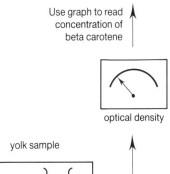

Figure 69 How a spectrophotometer is used to measure pigment concentration

(ii) Yolk colour and the reflection of light.

Yolk colour can be determined by measuring the amount of light which is reflected from the surface of the yolk. This can be determined by using a colour reflectance attachment linked to the spectrophotometer. The light reflected by the yolk at a particular wavelength can be used to determine pigment concentration when compared with standard solutions.

7.3.5 Common yolk defects

The quality of the egg yolk may also be affected by one of several common defects:

- **mottled yolks** — spots or blotches on the surface of the yolk. This condition is more common in eggs which have been stored, especially if the temperature of storage has been high. It is due to the movement of albumen into the yolk.
- **blood spots** — these may develop within the yolk as a result of the rupture of small blood vessels in the follicle within which the yolk develops when it is attached to the ovary.

7.3.6 The composition and structure of albumen

Albumen or egg white is made up of four layers, the proportions of each will vary with the breed of the hen, egg size and environmental conditions as shown in table 7.8.

Table 7.8 Mean proportions of different albumens within the hen's egg

ALBUMEN	MEAN %	RANGE
Outer thin	23	10–60
Thick	57	30–80
Inner thin	17	1–40
Chalazae	2	2

Albumen is composed mainly of water and a mixture of proteins. The protein **ovomucin** forms flexible microscopic fibres which produce the gel-like structure of the albumen. There is four times as much ovomucin in thick albumen than in thin albumen. These fibres are surrounded by an aqueous solution of other proteins which are responsible for the functional properties of the egg white:

- **ovalbumin** – denatures on heating so that egg white goes solid when cooked.
- **lysozyme** and **ovoglobulins** – produce foams so that egg white can be used in products such as meringue.

Table 7.9   The composition of albumen

COMPONENT	% OF ALBUMEN
Protein	9.7–10.6
Lipid	0.3
Carbohydrate	0.4–0.9
Ash	0.5–0.6

7.3.7 Albumen quality

Albumen quality or condition is important to the overall quality of the egg. Once the egg is cracked, a firm albumen gives the egg its structure. The condition of the albumen is determined by the proportion of thin white to thick white within the egg. An egg containing a large proportion of thin white will have a flatter albumen on cracking and be of lower quality. The proportion of thick to thin albumen is affected by factors both before and after laying. The breed, age and general health of the hen all influence the ratio of thick to thin albumen. After laying, the environmental conditions under which the eggs are stored are important. As eggs are stored, the thick albumen begins to thin. The rate of thinning is influenced by factors such as temperature and humidity. This is discussed in a later section.

Albumen condition can be assessed in several ways. The most commonly used methods are **candling** and the measurement of **Haugh units**.

(i) **Candling** Egg candling is used to assess albumen quality in unbroken eggs. Candling relies on the fact that the eggshell and its membranes are able to diffuse light and reflect it internally. This allows the inside of the egg to be seen. Albumen condition can be assessed by observing the intensity of the yolk shadow within the egg as it is rotated in front of a light source. If the albumen is firm, the yolk will not move and only a faint shadow is seen. If, however, the albumen is of low quality, the yolk is able to move when the egg is rotated and the yolk shadow will vary.

This method of assessment is fairly crude but can be performed rapidly on large samples of eggs without breaking the shells. It is, therefore, a very useful technique and is used routinely at commercial grading and packing stations.

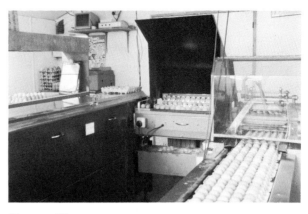

Egg candling

(ii) **Haugh Units** In 1937 Haugh described a method of determining albumen quality of opened eggs. The measurement of Haugh units is still used today. The egg is cracked onto a flat surface and the height of the albumen measured using a micrometer as shown in figure 70. The Haugh unit is calculated using the equation shown below and takes into consideration the albumen height and egg mass.

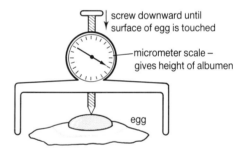

Figure 70 Using a micrometer to measure albumen height

Haugh Unit = $100 \log (H + 7.57 - 1.7W^{0.37})$
(HU)

H = height of albumen
W = mass of egg (g)

Calculating Haugh units

Haugh units give a good measure of albumen condition. The higher the unit, the better the quality of the albumen. However, the process is time consuming and so cannot be performed as a routine assessment. Care must be taken to ensure that albumen is not punctured on cracking the egg and albumen height must be measured directly after breaking.

Now try Investigation 12 Investigating Egg White in *Animal Science in Action Investigations*.

7.3.8 The composition and structure of the shell and its membranes

The egg shell acts as a package for the egg itself. It is porous, allowing materials to diffuse in and out of the egg, and strong, so provides protection.

The shell is lined by two protein membranes which lie between the albumen and the shell itself. The inner membrane is 0.22 μm thick and consists of three layers of protein fibres which run parallel to the shell and at right angles to each other. The outer membrane is thicker than the inner and consists of six layers of fibres. The outer membrane is attached to the shell by cones which extend from the shell into the membrane.

The shell consists of a matrix of interwoven protein fibres which are impregnated with calcite crystals. These crystals are composed mainly of calcium carbonate (98.2%) with small amounts of magnesium (0.9%) and phosphate (0.9%). The egg shell is produced by the uterus. Initially, calcite crystals are deposited on to the shell membrane to form the inner shell (the maxillary layer). This layer is spongy in appearance and is soon covered by columns of hard calcite crystals. The addition of magnesium increases the strength of this layer. As the calcite crystals grow, they leave between 7000–17 000 pores at right angles to the shell surface. These allow materials to diffuse through the shell to or from the yolk and albumen. The shell is covered with an insoluble protein cuticle which helps prevent microbial attack.

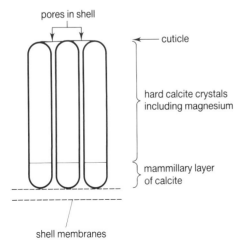

Figure 71 The structure of the shell

7.3.9 Shell colour and quality

Shell colour is important to consumers; in the UK most prefer brown eggs. The colour of the shell is determined by the presence of a pigment called **proporphyrin**. The production of this pigment is genetically controlled and so the colour of the egg shell will depend on the breed of the hen from which it is obtained.

Shell colour can be assessed subjectively by comparison with standards, however objective measurement of shell colour is also possible using photoelectric reflectance techniques. If a light is shone at a brown egg, it will reflect less light from its surface than a white egg. The darker the egg, the less light will be reflected. The percentage of light reflected from shells can be measured and the measurements used to grade eggs. Shell colour can be assessed commercially at a rate of 600 eggs per hour.

Table 7.10 Assessing shell colour

EGG SHELL COLOUR	% REFLECTANCE
White standard	100
White egg	80
Tinted egg	40–60
Brown egg	30

The **porosity** of the shell plays an important part in determining egg quality. Pores in the shell surface allow materials to be exchanged within the egg. This is important if the egg contains a growing embryo, but in commercial egg production high shell porosity is undesirable as it speeds up deterioration of the egg due to the loss of water and carbon dioxide and the entry of micro-organisms.

Shell porosity can be determined in a number of ways:

- Direct counting by staining sections of the shell and counting the number of pores per unit area.
- Measuring the rate of gas flow through the shell. This can be investigated by attaching a syringe to the air space end of the egg. The syringe contains a standard volume of atmospheric air which will be exchanged through the shell pores to the embryo inside the egg. After a set period of time, the air within the syringe is analysed and the rate of oxygen uptake can be calculated. This can be related to the porosity of the shell, however,

the metabolic rate of the embryo will vary depending on its stage of development and so this method may only be useful when comparing the porosity of eggs of the same age.

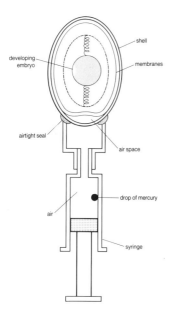

Figure 72 Measuring rate of gas exchange in eggs

- Measuring loss of mass under controlled conditions. This is the most favoured method of assessing porosity as it is non-destructive. The mass of the egg is determined before and after storage for 14 days at 38°C and 80% relative humidity, although shorter storage periods can also give reasonably reliable results.

> **Now try Investigation 14 Measuring Shell Porosity in *Animal Science in Action Investigations*.**

The shell provides a mechanical barrier which protects the internal egg. If this barrier is to be effective, it must be strong. Mechanical damage accounts for losses of about 9% during the egg production process as shown in table 7.11.

Table 7.11   Losses during egg production

STAGE OF PROCESSING	% CRACKED
Leaving farm	2.8
During grading	2.6
Between grading and reaching wholesaler	3.4
Total	8.8

Shell strength can be measured directly or indirectly. Direct measurements involve the measurement of how the shell resists internal and external pressure. Indirect measurements involve the measurement of shell thickness and relate this to strength. These methods are summarised in figure 73.

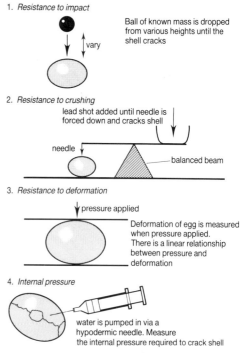

A. Direct measurement of shell strength

1. *Resistance to impact*

   Ball of known mass is dropped from various heights until the shell cracks

2. *Resistance to crushing*

   lead shot added until needle is forced down and cracks shell

   needle   balanced beam

3. *Resistance to deformation*

   pressure applied

   Deformation of egg is measured when pressure applied. There is a linear relationship between pressure and deformation

4. *Internal pressure*

   water is pumped in via a hypodermic needle. Measure the internal pressure required to crack shell

B. Indirect measurement of shell strength

1. *Direct measurement* of shell thickness using a micrometer
2. *Beta scattering* instruments measure shell thickness without breaking the egg
3. *Specific gravity* of a newly laid egg can be related directly to shell thickness. Eggs can be immersed in a test solution of a known specific gravity – those which float are suitable for sale.

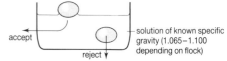

accept   reject   solution of known specific gravity (1.065–1.100 depending on flock)

Figure 73 Measuring shell strength

> **Now try Investigation 13 Measuring Shell Strength in *Animal Science in Action Investigations*.**

7.3.10 Changes in egg quality on storage

When eggs are stored their quality will deteriorate. This is mainly due to the movement of water through the layers of the egg and the loss of carbon dioxide through the shell.

As an egg is stored, the loss of carbon dioxide by diffusion through the pores of the shell causes the pH of the albumen to rise dramatically. At 20°C the pH of the albumen may rise from 7.6 to

9.4 in a few days. Albumen is composed mainly of protein and its condition is greatly affected by a rise in pH. This results in thinning of the thick albumen and a decrease in functionality of the albumen proteins. This means that the white of stored eggs will not produce foams or coagulate on heating as rapidly as newly laid eggs. The effect of temperature on albumen condition is shown below.

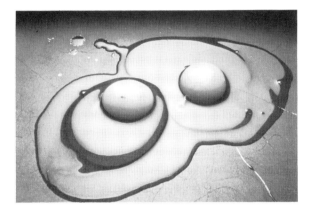

The effect of storage temperature on albumen condition

The rise in pH also causes the vitelline membrane to become weakened, allowing water to diffuse in to the yolk. This causes the yolk to become flatter and makes it difficult to break the egg without damaging the yolk.

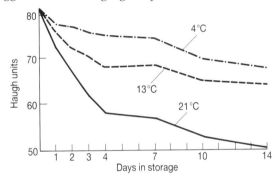

Figure 74 The effects of storage on eggs - yolk size

7.4 SYSTEMS OF MANAGING EGG PRODUCTION

The last 30–40 years have seen great changes in methods of keeping poultry. Traditionally, small flocks were kept on many farms. These were "free-range" obtaining much of their food from the environment. As recently as 1960 over 30% of laying poultry were free-range. By 1984 this figure

had dropped to 1.9% and although in the late 1980s there was a swing back towards free-range poultry and egg production, today the vast majority of eggs are produced intensively.

Table 7.12   Systems of egg production

SYSTEM OF MANAGEMENT	% LAYING BIRDS		
	1960s	1970s	1980s
Free-range	30.9	6.1	1.9
Battery	19.3	85.0	96.1
Others	49.8	8.9	2.0

Intensive poultry farming has overcome many of the problems of traditional free-range farming and led to an increase in productivity of birds. Egg laying is controlled by day-length as discussed earlier in this chapter. It is possible to control light duration and intensity in intensive systems and this coupled with warmer temperatures has resulted in higher productivity.

Table 7.13   Productivity of hens kept under different systems of management

SYSTEM OF MANAGEMENT	EGG YIELD per hen/per year
Free-range	192.1
Battery	245.4
Others	224.5

7.4.1 Free-range eggs

Only a small proportion of eggs are produced in this way. This system requires less capital investment than other systems but more space and a much higher input of labour. Free-range eggs are often sold at a premium price and so this system of production is often profitable if sufficient land is available.

The EC have laid down minimum requirements for hens producing free-range eggs. These include:

- Hens must have continuous daytime access to open air runs.
- Ground must be mainly covered with vegetation.
- A maximum of 1000 hens per hectare (1 per 10 m^2).
- Shelter must contain perches which allow 15 cm for each hen and at least one third of the floor must be covered with straw or a similar material. There must be space for droppings to be collected.

Birds are often kept in moveable houses which are mounted on wheels and hold up to 50 birds. These may have a slatted floor to allow droppings to be collected. It is not possible to control environmental conditions when using this system and so birds are often exposed to winter weather. Hens will require a higher level of feeding to compensate for the lower temperatures experienced inside the house due to the lack of temperature control.

Free-range egg production

7.4.2 Egg production in laying cages

The use of "batteries" of laying cages for keeping laying birds is the most popular system of commercial egg production. Over 90% of all eggs produced in the UK are produced in this way. The main advantages that this system has to the producer is that large numbers of birds can be kept in a relatively small space which means that labour costs are low and increased mechanisation is possible. It is also possible to control environmental factors such as temperature and day-length.

Laying cages may house 2–25 birds. They have sloping wire floors so that when eggs are laid they roll forward out of reach of the hens. Droppings pass through the wire flooring to a belt or pit below for collection. Cages are usually 450 mm deep and 450 mm high at the front, sloping to 350 mm at the back. Feeding troughs are fitted to the front of the cage. Each bird must have 100 mm of trough space.

Cages can be arranged in several ways:

- Stacked. Cages arranged three or four tiers high are mechanically or hand cleaned.

Stacked laying cages

- Stair step (deep pit). Cages are staggered so that droppings fall into a large pit. These are cleaned out at the end of the laying cycle.

Stair step laying cages

- Flat deck single tier of cages suspended over a pit for the collection of droppings.

Flat deck laying cages

7.5 CHICK PRODUCTION

7.5.1 Fertilisation of the egg

If chicks are to be produced, the eggs laid must have been fertilised. After mating, the sperm of a chicken is stored in the folds of the oviduct. Chicken sperm is viable for up to 32 days. As the yolk enters the infundibulum, the sperm are released from the walls of the oviduct and fertilisation will occur in the germinal disc (the blastodisc) at the top of the yolk. The embryo will develop around the blastodisc and by 48 hours a well defined system of blood vessels have developed in order to obtain nutrients from the yolk.

7.5.2 Embryonic development

Day One

The entry of the sperm stimulates cell division in the blastodisc so that a layer of cells (the **blastoderm**) is formed rapidly. Below the blastoderm there is a space called the **subgerminal cavity** which is in contact with the yolk. The blastoderm is divided into the **area pellucida** cells which have no contact with the yolk, and the **area opaca** cells which touch the yolk at each end of the blastoderm. These cells digest yolk so that nutrients are available to the rest of the blastoderm (see figure 74). This stage is called the **blastula**. During **gastrulation**, the cells of the embryo begin to differentiate and organise themselves, and the organs begin to develop. Cells sink inwards and form a layer along the subgerminal cavity in contact with the yolk. These cells form the **endoderm** which will develop into the internal organs such as the intestines and lungs. The egg is usually laid at this stage. The mesoderm which will develop into the bones, muscles, blood, excretory and reproductive systems has begun to organise itself, as has the ectoderm which will form the nervous system, skin and feathers (see figure 75).

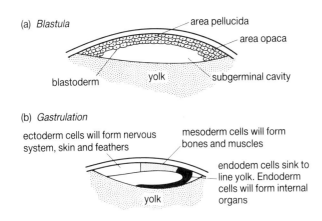

Figure 75 Chick development – day one

Day Two

The mesoderm begins to develop. The mesoderm cells are found along the top of the subgerminal cavity. These cells move forward forming a ridge called the **primitive streak**. The **primitive pit** develops at the front of this, and behind it cells pile up to form the **primitive knot**. The primitive knot will develop into the notochord. The

mesoderm cells sink through the **primitive groove** which appears along the middle of the primitive streak and to lie on top of the endoderm. Islands of blood develop within this area. These develop into the **vitelline blood vessels** which carry digested food to the growing embryo. This is shown in figure 76.

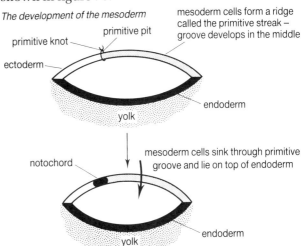

Figure 76 Chick development – day two

Day Three

The neural plate folds to form the neural tube from which the brain and spinal cord will develop. At the front of the embryo, the ectoderm folds to form the head fold and the endoderm folds to form the foregut. The heart is formed and starts to function but is situated outside the body.

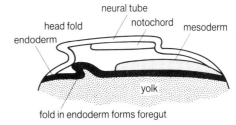

Figure 77 Chick development – day three

Day Four

Leg and wing buds and the tail can be seen. The nervous system and the brain are developing as are most of the other major organ systems.

Day Eight

Feather tracts can be seen on the skin. The wings and legs are developed. The heart is incorporated into the chick's body. Circulation of materials from the yolk sac now occurs through the umbilicus.

Day Sixteen

The bones have become calcified, the beak and claws are developed. The volume of amniotic fluid begins to decrease as the chick becomes ready to hatch.

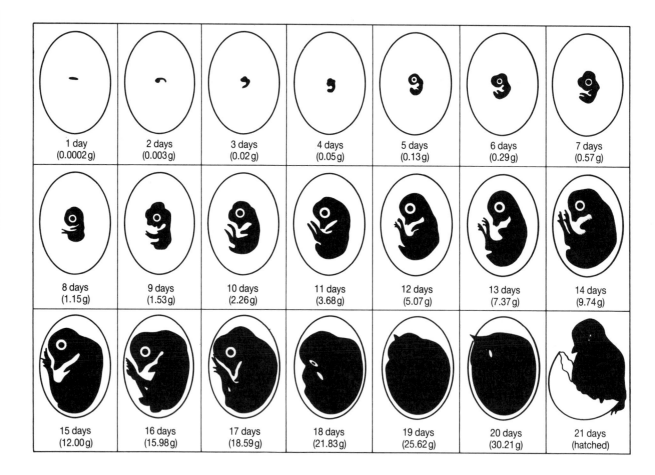

Figure 78 Summary of embryonic development in the chick

7.5.3 Extra-embryonic membranes

The developing embryo is surrounded by a series of extra-embryonic membranes — the yolk sac, amnion, chorion and allantois.

The **yolk sac** begins to form before the egg is laid by the spreading of the endoderm and ectoderm over the yolk surface. The yolk sac is responsible for digesting materials in the yolk. This food is then transported via the blood vessels to the growing embryo. The yolk sac is absorbed into the gut just before hatching, where it degenerates.

The **amnion** develops during the second day after laying. It forms a cavity which surrounds the embryo. This fills with amniotic fluid which bathes the embryo preventing dehydration and protecting it from mechanical damage or harm due to temperature changes.

The **chorion** surrounds all of the other extra-embryonic membranes. It develops at the same time as the amnion but is in contact with the albumen. As the albumen is used up by the embryo, the chorion fuses with the **allantois** and eventually becomes pressed against the shell membranes. In this position, it plays an important part in the absorption of oxygen across the shell.

The **allantois** grows out from the hind gut rapidly to fill the chorionic cavity. The allantois has a highly developed circulatory system and functions as an extra-embryonic lung until the chick hatches. It also stores uric acid which is produced by the kidneys as a by-product of metabolism.

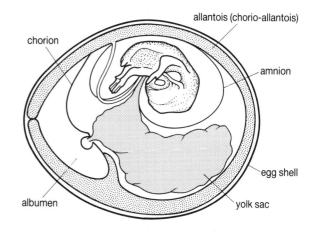

Figure 79 The extra-embryonic membranes

7.5.4 Hatching

A few days before the chick is due to hatch, the yolk sac contracts and is absorbed into the body. Each time the yolk sac contracts, the muscles in the back and the neck twitch. This causes the beak to be jerked upwards so that it eventually makes contact with, and ruptures, the allantois and the shell membranes at the air space end. The chick will begin to breath and within 5 or 6 hours the shell will be cracked. About 24 hours after the chick begins to breath, it will finally emerge, wet and exhausted from the shell.

Hatching

7.5.5 The storage of eggs for hatching

Commercially, most eggs are hatched in artificial incubators. Fertilised eggs are often stored before incubation, in some cases for several days. Storage conditions are very important if the eggs are to remain viable. The recommended conditions for storage are shown in table 7.14.

Table 7.14   Recommended storage conditions prior to incubation

STORAGE TIME (days)	TEMPERATURE (°C)	RELATIVE HUMIDITY %
1–3	20	75
4–7	13–16	75
8–14	11–12	75 (enclosed in plastic bags)

Ideally, eggs should not be stored for more than seven days as the percentage hatchability of the eggs will fall. To improve the hatchability of eggs after prolonged storage, eggs should be sealed in plastic film and surrounded by an atmosphere of nitrogen.

7.5.6 Incubation of eggs

Commercial incubators are able to hold up to 100 000 eggs. The developing embryo is extremely sensitive to temperature. If an egg is incubated naturally by a hen, the centre of the egg reaches a temperature of about 37.8°C and the surface between 39.2–39.4°C. Most incubators maintain a temperature of between 37.5–37.8°C to ensure that the centre of the egg reaches the optimum temperature for embryo growth. The embryo itself will generate heat as it develops and so most commercial hatcheries lower the temperature of the incubator during the last two days in order to avoid late embryo mortality.

The relative humidity inside the incubator can vary more than the temperature before hatchability is affected. The optimum relative humidity for incubation is about 60% (i.e. 2.75 kg water in 100 m^3 of air)

During incubation eggs should be either laid flat or with the broad end upper most. This allows the developing embryo maximum access to the air space. Eggs must be turned regularly to prevent the embryo from becoming stuck to any of the other structures within the egg. This is especially important during the early stages of development. Eggs may be turned either automatically or by hand.

QUESTIONS

1 Albumen quality or condition is very important to the overall quality of the whole egg.
   a) What factors within the egg determine the quality of the albumen in a newly laid egg?
   b) As eggs are stored, the pH of the albumen rises from 7.6 to, in some cases, as much as 9.4. What causes this increase in pH?
   c) Albumen is composed mainly of protein. Explain how the change in pH described above affects these proteins? What is the overall effect of this on albumen condition?
   d) Prolonged storage will also affect the quality of the yolk. Why does this occur?

2 The table below shows changes in egg production between the mid-1970s and 1980s.

	1974	1984
Numbers of laying birds (millions)	50.84	41.14
Egg production (millions)	1143	1005

a) Calculate the percentage decrease in both the numbers of laying birds and the numbers of eggs produced between 1974 and 1984.

b) Explain why egg production has not fallen as rapidly as the numbers of laying birds. What factors are likely to be responsible for this?

c) The table below shows changes in the consumption of eggs in the USA since the 1950s. Suggest three reasons for this downward trend in consumption.

YEAR	PER CAPITA CONSUMPTION OF EGGS IN THE USA (approx.)
1950s	390
1960s	300
1980s	244

3 The colour of the egg yolk is very important to both the consumer and the food industry. Yolk colour can be manipulated by the producer by the selection of particular types of poultry feed. Imagine you are a poultry farmer producing eggs for a large supermarket chain. It is important that the colour of the yolks in your eggs remains the same in every batch so that they are suitable for sale to the supermarket. List the method that could be used to determine yolk colour and assess the suitability of each method for this type of "on farm" analysis.

4 The egg production of a flock of free-range chickens was monitored over the course of a year and is shown in the table below.

MONTH	MEAN EGG PRODUCTION PER DAY
January	40
February	53
March	70
April	80
May	79
June	82
July	82
August	79
September	65
October	52
November	47
December	40

a) Draw a graph to illustrate this data.

b) Explain why egg production is not constant throughout the year.

c) Suggest a method by which the producer could ensure a constant supply of eggs throughout the year. Sketch a line on your graph to show the predicted effects of these measures.

5 The majority of eggs for the UK market are produced intensively by so called "battery hens", however since the 1980s there has been renewed interest in the production of "free-range" eggs using more traditional methods. Compare these two techniques, with respect to the points listed below:

a) Productivity.

b) Quality of product.

c) Consistency of supply throughout the year.

d) Animal welfare.

e) Disease management within the flock.

Write a brief report for the school magazine which expresses your views on battery farming. Try to back up your arguments with as many facts as possible.

6 Microscopic examination of the shell of a bird's egg reveals the presence of a large number of pores through which gas exchange takes place.

a) Suggest four factors, other than the number of pores, that would influence the rate of diffusion of gases through the egg shell.

b) Assuming that the pores are evenly distributed, and that the egg shell is the same thickness all over, suggest a method for estimating the total number of pores.

c) Explain how the survival of the developing chick would be threatened if the permeability of the shell was too great.

In order to investigate the composition of the gas within the air space of developing eggs without damaging the shell itself, a syringe containing a standard volume of atmospheric air and a drop of mercury was connected to the shell by an airtight seal as shown in the diagram below.

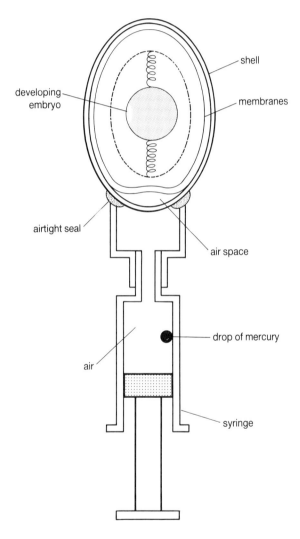

Apparatus to measure composition of air inside air sac

After the apparatus had been set up for six hours, the syringe, while still attached to the egg, was lifted to a vertical position and then disconnected from the egg. The air within it was the analysed.

d) Suggest the reason for including the drop of mercury in the syringe.

e) (i) What assumption were the experimenters making about the composition of the gas in the syringe after six hours?

(ii) How could the investigation be extended to determine the accuracy of this assumption?

f) The rate of oxygen uptake by the developing chick was found to increase considerably during the course of incubation. Assuming that there was no change in the structure of the shell, give two factors that could increase the rate of oxygen uptake as the embryo developed.

JMB Nuffield

7 The exchange of gases between the developing chick embryo and the surrounding atmosphere takes place through pores in the shell. Incubation, the time taken from laying to hatching, for a hen's egg takes about 21 days.

a) How, and for what reason, would you expect the diffusion gradient of oxygen across the shell to change during the incubation period?

b) At an altitude of 3800 m, reduced atmospheric pressure results in a low partial pressure of oxygen and an increase in the rate of diffusion of gases. Explain how these two features might combine to affect the development of the hen's egg laid at altitude.

c) The eggs of birds that live naturally at high altitudes tend to be small in relation to body size and have a reduced number of pores per unit area of shell. Suggest how the characteristics described may adapt the eggs of these birds to successful development at high altitude.

AEB 1992

BIBLIOGRAPHY

Brooks, J., Taylor, D.J. (1977) *Eggs and Egg Products*. HMSO.

Hamilton, R.M.G. (1982) *Methods and Factors That Affect Egg Shell Quality*.

Solomon, S.E. (1991) *Eggs and Eggshell Quality*. Wolfe.

Stadelman, W.J., Coterill, O.J. (1986) *Egg Science and Technology*. AVI Publishing Company Inc.

Vadehra, D.V. (1989) *The Avian Egg*. Wiley Interscience.

8 FISH PRODUCTION

LEARNING OUTCOMES

After studying this chapter you should be able to:

- distinguish between demersal and pelagic fish,
- talk about the advantages and disadvantages of different fishing methods,
- talk about the problem of overfishing and list and evaluate methods which could be employed to reduce the problem,
- compare the popularity of fish-farming throughout the world,
- explain the importance of controlling the environment within which fish are reared, with respect to oxygen levels, temperature, salinity and acidity,
- outline the processes of gamete production, spawning, fertilisation and hatching in the fish and explain how human intervention has manipulated these processes,
- list and describe the methods available for the "growing on" of farmed fish.

8.1 FISH FOR FOOD

As the world's population continues to expand, more pressure is put on agricultural land in an attempt to increase the production of both crops and animals for meat. Fish is a valuable human food source which is rich in protein and low in fat. In recent years there has been increased demand for fish and this is likely to continue. Most of the fish we consume comes from the sea, however **aquaculture** or fish-farming is becoming increasingly more popular.

There are many fish species used as food worldwide. Some live in fresh water, some in salt. Species of fish commonly used as food in the UK include the sea-fish cod, haddock, mackerel, tuna, sardine and herring, and the fresh-water fish trout and salmon. These are all bony fish or **teleosts**. This means that their skeleton is composed of bone unlike the **cartilaginous** fish, for example dogfish, sharks and rays, which have skeletons made up of cartilage.

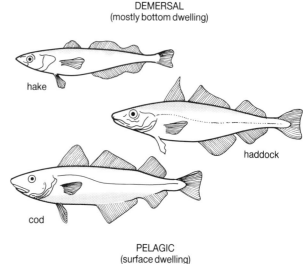

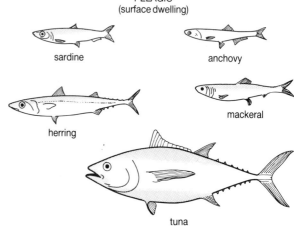

Figure 80 Some common fish used for food

Fish can be described as either **demersal** or **pelagic**. Demersal or white fish have a large proportion of white muscle. They are flat and tend to swim close to the sea or river bed. Pelagic or oily fish are rounded in shape and are able to sustain high swimming speeds for long periods of time due to the larger amount of lipid their bodies contain. In general, fish contain a high percentage of protein (up to 22% and have a low fat content. The chemical composition of some common fish species is shown in table 8.1 below.

Table 8.1   The chemical composition of some common fish

SPECIES	WATER %	FAT %	PROTEIN %
Cod	80–83	0.1–1.0	15–19
Halibut	75–79	0.5–10	18–19
Herring	60–80	1.0–25	18
Salmon	67	0.4–14	22

Fish is also rich in minerals, especially iron and sulphur, and low in cholesterol.

8.2   THE SEA-FISHING INDUSTRY

In the UK we catch about 800 000 tonnes of fish from the sea each year. The majority of this being mackerel, cod and haddock. Worldwide, only a few of the 20 000 species of fish which exist are commonly used for food. This is shown in table 8.2. below. This situation has resulted in the over-exploitation of some traditional fish species. Careful management is required if overfishing is to be prevented and stocks of these fish species are to be maintained.

Table 8.2   Main fish species caught worldwide in 1984

SPECIES	CATCH ($\times 10^6$ tonnes)
Pollack	5.99
Pilchard	10.18
Capelin	2.58
Mackerel	4.62
Cod	1.97
Herring	1.19
Tuna	1.05
Sardine	0.88

8.3   FISHING METHODS

There are many methods of catching sea-fish. The method employed will depend on factors such as population size, species and the eventual quality of the fish removed. Commercial techniques are mainly based on either netting or hooking.

8.3.1   Netting methods

The most commonly used netting techniques are:

- trawling,
- seine nets,
- drift nets.

Both midwater and bottom **trawling** are commonly used to catch most of the important fish species. Large nets are dragged either through the water or along the sea bed (see below). Trawling is extremely efficient at catching fish – one trawl may land up to 36 tonnes – however there are many problems associated with this method.

Trawler

- Trawling nets become very heavy when full and so large boats are needed to pull them. This means that trawling also requires large amounts of fuel.
- The size of the net mesh is supposed to prevent very small fish from being caught. However, as the nets fill up, the holes in the mesh become blocked and so the small fish cannot escape.
- The volume of fish caught per trawl makes rapid processing difficult and so it is difficult to produce very high quality fish using this method.

Seine nets are used to catch schools of fish, for example mackerel. The net is towed to the school of fish by two small boats. The school is then surrounded by the net and the bottom of the net is sealed by pulling a rope. The net is tightened so that the fish are held within a small area from which they can be landed by the ship's crew using hand nets. This process is shown in figure 81.

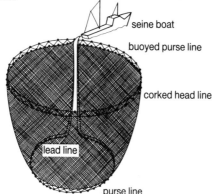

Figure 81 Seine netting

Drift nets are nets which hang in the water suspended by buoys. The nets are often set in areas where fish are known to migrate. The fish cannot see the nets and so swim into them getting their gills caught. Drift nets were extensively used before the Second World War, however, they are less popular nowadays for two reasons:

- the fish which are caught struggle to get free and so become bruised. This means that they produce a lower quality product,
- the nets are not selective and so catch a wide variety of fish as well as marine mammals, such as dolphins.

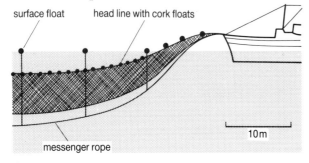

Figure 82 Drift nets

8.3.2 Hooking methods

Hook and line fishing can be used to catch many common fish species including tuna and mackerel. This traditional method is not extensively used nowadays, however, as it is more labour intensive than netting techniques. Hooks are attached to lines which are suspended from buoys in the water. Lines, which are often baited, can be over 6 km in length.

8.4 THE MANAGEMENT OF SEA-FISHING

If the seas are to continue to provide us with fish for food, it is essential that fishing is managed so that stocks are not depleted. In order to ensure that overfishing does not occur, it is important to know the maximum yield of each fish species which can be harvested per year without adversely affecting future harvests.

8.4.1 The population dynamics of sea-fish species

As with all populations, the rate of growth of fish populations, under normal conditions, will follow a **sigmoidal** pattern. The rate of growth of the population can be expressed in terms of the rate at which the **biomass** is increasing. This will take into account both increases due to growth of existing fish, the birth rate and the death rate. When the birth and growth rates balance the losses due to death, an equilibrium will exist (see figure 83).

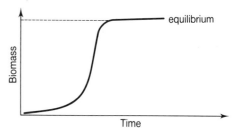

Figure 83 Sigmoidal shape of the growth curve for a population

If fish stocks are to be maintained, fishing must only harvest stocks at a level which allows the biomass to return quickly to this equilibrium point. Overfishing may result in the fish population falling to levels from which it is unable to recover. The effects of overfishing on the North Atlantic herring in the 1970s are shown in figure 84.

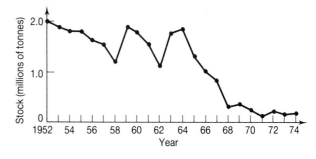

Figure 84 The effects of overfishing on the herring population in the 1970s

Large catches of herring landed in the 1950s and 1960s led to the near depletion of the herring population in the North Atlantic. In 1976, herring fishing was banned by the International Council for the Exploration of the Sea. The ban was lifted in 1983, but herring fishing is now tightly controlled by the EC. The ban had devastating effects on some fishing ports and the herring processing industry.

8.4.2 Preventing overfishing

Fishing can be controlled by a series of regulatory measures. Many of these are controversial as they aim to limit the yield of fish harvested and this will have economic implications for those within the fishing industry.

- **Regulating mesh size** This measure aims to control the age of the fish caught. Ideally the mesh size is set so that small fish are able to pass through. These fish are then able to grow and spawn, thus adding to the population, before they are harvested.
- **Setting fishing quotas** In 1984 the EC lifted the ban on herring fishing in the Northern Atlantic and introduced an annual quota. Fishermen were allowed to harvest 84 300 tonnes of herring only. Fishing quotas are difficult to control and often lead to large numbers of small fish being caught and then returned to the sea, whilst the larger, more marketable fish are kept. These small fish are usually dead or dying and so do not grow and reproduce. This means that large numbers of fish are wasted.
- **Controlling fishing areas** In 1977 the EC adopted a 320 km exclusion zone around their waters in an attempt to prevent fishing by non-EC countries. The enforcement of the exclusion zones is extremely difficult and often leads to confrontation.
- **Reducing fleet sizes** In 1987 the EC issued guidelines which aimed to encourage member states to reduce their fishing fleets by 3%. This measure has been fairly unsuccessful, especially in the UK where the fishing fleet is currently 30% too large! The EC are encouraging governments to pay grants to fishermen who stop fishing, however, the British government is reluctant to implement such a scheme.

8.5 FISH-FARMING

One way of overcoming the problems of overfishing is by artificially rearing fish. Fish-farming or **aquaculture** mostly involves the production of fresh water fish. It dates back to

2000 BC where there is evidence of carp production. In medieval times carp were used to stock moats and ponds so that fresh fish could be obtained throughout the year. Fish-farms were common in England and Europe up until the 19th century when the availability of fish from the sea led to their decline.

Marine fish-farming is more problematic because it is difficult to recreate the sea environment required. Most successful marine fish-farms produce shellfish such as mussels and oysters rather than fish. As shellfish are not very mobile, they can be produced in areas where the sea has been enclosed.

Worldwide, the annual production level of fish for human consumption is about 50 million tonnes. Only about 10% of this is produced by aquaculture. The United Nations aim to increase this figure to 50% over the next twenty years, thus reducing the problems of overfishing. Many species are suitable for fish-farming. Some commonly used species are shown in figure 85.

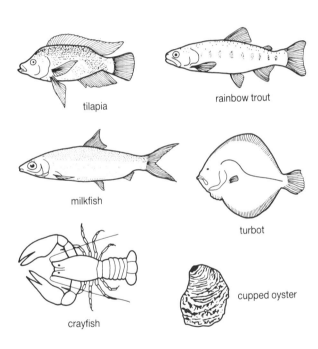

Figure 85 Species commonly used in fish farming

Table 8.4 World production of fish and shellfish by aquaculture

	APPROXIMATE YIELD ($\times 10^6$ tonnes)
Fish	4.0
Seaweed	1.6
Shellfish (mainly oysters)	1.0
Total	6.6

There are about 400 fish-farms in Great Britain, producing about 10 000 tonnes of fish a year. Most of the fish produced belong to the **salmonid** family – about 1200 tonnes of Atlantic salmon and 8000 tonnes of rainbow trout, although other species such as eels, carp and marine fish are also produced in smaller quantities. Table 8.5 shows some of the producers of farmed fish worldwide and the diversity of the species they produce.

Table 8.5   Species produced by aquaculture in selected areas

COUNTRY	ESTIMATED PRODUCTION 1983 ($\times 10^6$ tonnes)	PRINCIPAL SPECIES
China	1.5	carp, tilapia
CIS	0.3	carp, sturgeon, whitefish
UK	0.1	rainbow trout, Atlantic salmon

8.6   FISH-FARMING AND THE AQUATIC ENVIRONMENT

Successful aquaculture relies heavily on the quality of the water in which the fish are kept. Water supplies the fish with oxygen and maintains the fish's temperature as they are cold blooded (**poikilothermic**). Other parameters such as salinity and pH are also important. A producer must maintain optimum levels of each of these factors if maximum productivity is to be achieved.

8.6.1   Oxygen level

Salmon and trout require an oxygen level of 9–12 mg of oxygen per litre of water to maintain their metabolism. The amount of oxygen which can be dissolved in water depends on the temperature of the water (see later), the pressure and the salts which are dissolved in it. Seawater can dissolve less oxygen than fresh water due to the amount of salts which are present (8.1 mg/l in seawater compared with 10.2 mg/l in fresh water).

As oxygen is continually removed from the water by the fish, it must be continually replaced. This is most effectively done by maintaining a continual flow of water either naturally by means of currents, or artificially by dams or pumps. If flow is interrupted, fish will die rapidly.

Many pollutants, especially organic matter, remove oxygen from water. This results in a Biological Oxygen Demand (BOD). If this becomes too large, the fish will suffocate and die immediately. Dead fish show signs of haemorrhages in the gills and skin. The appearance of dead fish with these symptoms in a farm tank should alert the producer immediately to increase the oxygen supply.

8.6.2   Temperature

The temperature of the water will determine the rate of growth, development and the amount of oxygen available.

Spawning and egg development will depend greatly on temperature. Optimum temperature for spawning will depend on the species. Fish grow faster at higher temperatures. The optimum temperature range for salmon and trout growth is 14–18°C. If fish are to be produced in open ponds or tanks, seasonal temperatures will limit growth rates. This means that production times will vary with geographical area, for example, trout may take 10 months to reach market size in southern England and up to 18 months in the cooler waters of northern England.

Oxygen becomes *less* soluble in water as the temperature increases. If fish are to be kept at high temperature, the water flow rate must also be increased to ensure that enough oxygen is supplied.

Table 8.6   The effect of temperature on the solubility of oxygen in water

TEMPERATURE °C	DISSOLVED OXYGEN (mg/1)
0	48.9
10	38.0
20	31.0
30	26.1
40	23.1

Now try Investigation 15 The Effects of Temperature on Fish Growth in *Animal Science in Action Investigations*.

8.6.3   Salinity and acidity

Many fish are able to tolerate varying levels of salinity. Fish such as salmon and trout live part of their life in freshwater and part in seawater under natural conditions, and so can tolerate increases in salinity once they reach a particular stage of development. Rainbow trout can, for example, be moved to salt water once they weigh 100 g, whereas Atlantic salmon must reach the smolt stage of development (about 1 year old) before they can tolerate salt.

The pH of the water is extremely important to the fish living in it. Salmon and trout prefer a pH of between 6.4 and 8.4. Factors which may alter this level include the build up of wastes within the tank or pond, pollution or the leaching of minerals from the soil. Variations outside this range can be very damaging.

8.7 INTENSIVE FISH-FARMING SYSTEMS

The main aim of the intensive fish-farm is to produce maximum yields of fish for maximum profit. Successful aquaculture relys on the ability of the producer to control the natural life cycle of the fish he wishes to cultivate. In the UK these fish are most commonly trout or salmon.

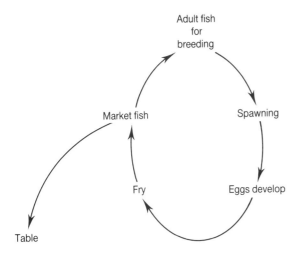

Figure 86 Outline of the fish life cycle

8.7.1 Gamete production, spawning and fertilisation

In the wild, spawning or gamete release is usually an annual event. The onset of spawning is often brought about by some external stimulus, such as temperature or day-length, although the fish must also have reached maturity before spawning can occur. This combination of external and internal factors cause the hypothalamus in the brain to produce releasing hormones which stimulate the gonadotrophic cells in the pituitary gland.

Gonadotrophic hormones are then released and these stimulate the ovary and testes to release the ova or sperm.

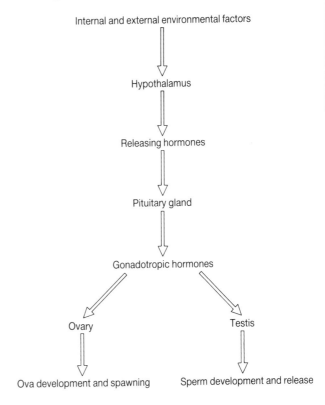

Figure 87 Gamete production in fish

Many species of fish will not spawn in captivity because the environmental conditions are not suitable. Many fish, including the salmonids only spawn once a year. The artificial inducing of spawning can be useful in both cases as it allows the production of fish for sale to be controlled.

Commercially, spawning is induced by hormone injection (**hypophysation**). This process was developed in the 1930s using hormones extracted from sheep and cattle to induce ovulation. However the technique now involves the use of hormones extracted from the pituitary glands of fish, preferably of the same species as the recipient. Hypophysation is only effective in inducing spawning if the recipient fish have reached maturity. The ova produced by the female must be in their final stages of development. If the injection is given too early, the ova fail to survive. Hypophysation can, therefore, only advance spawning by a few weeks and so this technique is really only useful as a method of synchronising spawning in intensive systems.

Once the fish are ready to spawn, the ova and sperm are often stripped from them, before they are released. This allows fertilisation to be performed under controlled conditions. About 12 hours after hypophysation is complete, the fish is anaethetised and the abdomen is massaged gently.

This should stimulate the release of the ova or sperm. The number of ova produced (fecundity) of the fish will vary with species as shown in table 8.7. In general, sperm from two males is added to the ova from one female.

Table 8.7 Fecundities of farmed fish

SPECIES	OVA PER KG BODY MASS
Atlantic salmon	1800
Brown trout	2100
Pink salmon	900
Rainbow trout	2200

The most commonly used method of fertilisation is dry fertilisation. This is summarised in figure 88.

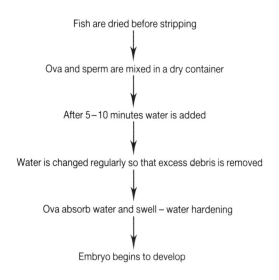

Fish are dried before stripping

↓

Ova and sperm are mixed in a dry container

↓

After 5–10 minutes water is added

↓

Water is changed regularly so that excess debris is removed

↓

Ova absorb water and swell – water hardening

↓

Embryo begins to develop

Figure 88 Dry fertilisation

8.7.2 Hatching

After fertilisation, the embryos are transferred to an incubator which is shaded from the light, where they should remain undisturbed for the next 10–15 days. Once the eye spots become visible (eyeing), the eggs are "shocked" by siphoning them into another container. This kills all damaged, unviable eggs so that they can be removed.

After hatching, the yolk sac remains attached to the small fish or fry. The developing fish uses the yolk sac for nutrition. As the fry mature, the yolk sac is absorbed and the fry are able to swim freely and search for food. At this stage, fry should be transferred to small holding tanks. Once the fish start to feed, it is important that they are fed

frequently. Fry will lose weight rapidly and die if feeding is not optimised. Fry should be fed for 20 hours a day from automatic feeders. At this stage in their life cycle, fry consume up to 10% of their body mass in food a day. The growth rate can be manipulated by the level of feeding. Most trout fry reach about 4–5 g after 130 days. At about 10 g they will be transferred to the main production tanks for growing-on.

Holding tanks for fry

8.7.3 Growing-on native fish species in the UK

In the UK the most commonly farmed native species are rainbow trout and the Atlantic salmon. Both of these species are farmed for food.

1 Trout

Trout are most often grown-on in fresh water pools, although they can survive in sea water. Once they reach a suitable size, trout are transferred to one of several systems:

- **Evacuated earthen ponds** are a cheap method of holding fish and are suitable so long as the soil type is such that the bottom of the pond seals itself so that water cannot drain away. If the soil type is not suitable, ponds can be lined with clay and/or butyl sheeting. Earth ponds are usually about 30 m long by 10 m wide by 1.5 m deep and are arranged around a central channel. Water flows into the ponds from a neighbouring river or stream and out into a central channel. Water is exchanged at a fairly slow rate (3–6 times a day) so fish stocking rates tend to be low (2 kg/m³). Fish are harvested by lowering the water level of the pond and pulling a net through the water. Fish may be placed in concrete holding tanks for a few days before they are sent to market in order to reduce the risk of muddy flavours!

Evacuated earthen ponds

- **Raceways** are long rectangular channels lined with concrete. They are 2–5 m wide and 0.5–2 m deep. High stocking densities (32 kg/m³ are possible because a higher rate of water flow is possible – water can be exchanged three times an hour. Each raceway can be individually drained for husbandry purposes or for harvest.

Concrete raceways used for trout production

- **Circular prefabricated tanks** are made of galvanised steel or fibreglass. Water flows in a circular fashion around the tank and drains into a central outlet. Tanks are about 4 m in diameter and 0.75 m deep. The circular water allows fish to be more evenly distributed throughout the water and also allows water to be self-cleaned.

Circular prefabricated tanks

The main market for rainbow trout is for fish weighing 170–280 g. It takes between 10–18 months for fish to reach this size from hatching.

2 Salmon

Salmon are transferred to sea cages for growing-on at about one year old. Sea cages are made of nets which are suspended from a floating framework and weighed down at the corners to produce a net cage with a volume of about 100–300 m³. Cages are moored offshore and hold up to a tonne of fish. Water can be exchanged rapidly through the cages by natural currents. The use of sea cages allows fish to be harvested easily and helps prevent the transmission of disease as fish can be isolated easily.

Sea cages

8.7.4 Growing-on carp

Carp are farmed extensively in China and Eastern Europe for food. There is little market for carp as food in the UK, but they are highly prized as ornamental fish and for coarse fishing.

Carp generally require higher temperatures for growing-on than native fish and so environmental control is essential. In most intensive systems, carp fry are imported from their native countries and placed in circular tanks, each containing 1000 gallons of water. The water within the tanks is maintained at 20–25°C by thermostat. The water is recycled throughout the system of tanks which hold the fish at various stages of development. Water leaves the tanks through a central drain which leads to the outlet pipe. This carries used water to the filtering system. Ammonia, nitrates and solids produced by the fish are removed before the water is aerated under pressure. The water leaves the filtration unit with an oxygen saturation of 110%. This level of saturation can support up to 120 kg of fish per tank. The stocking density of each tank will depend on the size of the fish. To ensure that the maximum stocking density is not exceeded, samples of fish from each tank are weighed weekly and fish removed from tanks as necessary. The rate at which water re-enters the tanks can be controlled by changing the diameter of the inlet tube with inserts.

Fish are fed automatically using pelleted food. At maximum growth rates carp can produce almost 1 kg of body weight per 1 kg of food consumed.

One of the main problems with this system of farming is that if the incoming fry are infected with any disease it can rapidly spread throughout the whole system. To help prevent this, incoming fry are treated with a mixture of chemicals to kill major fungal and bacterial infections which may be present. It is also essential that the producer knows the source of the fry being purchased.

Carp are usually sold on at about 3 kg in weight. They are acclimatised before leaving the fish farm so that they are able to survive in outside ornamental pools or freshwater lakes. Carp are sold on at about £8 per kilogram, however they can sell for more than £100 each at garden centres and aquariums.

QUESTIONS

1 The diagram shows a model for population regulation of a commercially important North Sea fish.

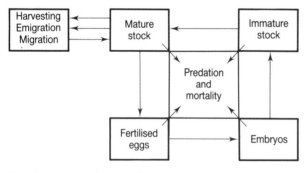

Population regulation of a commercially important fish

a) Under what circumstances will the population remain stable?
b) Explain why, in a managed population, it is desirable to keep the population in the log-phase of growth.
c) Fishing nets may catch fish of all sizes. How might this make it difficult to maintain a stable population when harvesting marine fish?

AEB 1992

2 Up to the 1960s, herring were very commonly found in the North Sea. Herring are traditionally caught by drift nets which are suspended in the sea by means of weights and floats. The herring do not see the nets and swim into them, getting caught by their gills. During October and November, mature herring migrate to the English Channel to spawn. Herring are often caught by drift netting in waters which are in their migration path. The table below shows the average number of herring of different ages caught per netting in 1955.

AGE OF HERRING (years)	MEAN NUMBER LANDED PER NETTING
3	8.63
4	6.61
5	3.71
6	2.02
7	0.73
8	0.65
9	0.40

a) Plot these data on to a histogram.
b) At what age do most herring appear to reach maturity?
c) Why is it favourable to catch herring shortly after they reach maturity and not before?
d) Nowadays drift netting is not commonly used. Suggest the reasons why it has become unpopular.

3 You are required to assess the current state of a fishery, and to propose measures for its most efficient management in the future.
a) (i) Describe in detail, one method that could be used to determine the age composition of the stock, and the rate of growth of the fish. Outline two other methods that might be used.
   (ii) Discuss the advantages and disadvantages of each of the three methods described.
b) What information would you need to enable you to decide whether or not the fishery was being overexploited?
c) How could overexploitation be prevented? What problems might arise from the use of the preventative measures you suggest?

JMB 1992

4 Successful aquaculture depends, in part, on the maintenance of a high oxygen level (9–12 mg/l) in the water.
a) Why is this important?
b) What environmental factors will affect the oxygen content of the water?
c) Explain how the oxygen content of water is maintained within intensive systems.

5 There has been some interest in the use of treated sewage to culture edible fish. Sewage is treated in the normal way and the resulting effluent is pumped into shallow pools where it is left to stand. This allows oxygention to occur

and for the establishment of a rich ecosystem containing aquatic animals such as plankton. Once this has been established, fish can be introduced into the pools for rearing.
a) Why is sewage effluent a good substrate for rearing fish?
b) Why do pools need to be shallow?
c) Why is it important to allow a rich population of aquatic animals such as plankton, to build up before fish are introduced?
d) Give two reasons why this use of sewage for fish culture is unlikely to become widespread.

6 Rainbow trout are commonly reared on fish-farms in the UK. Fry are produced in hatcheries usually, by stripping the eggs and sperm from the adult fish and mixing them *in vitro* to ensure fertilisation. The eggs are then incubated until the fry are ready to hatch. The table below shows the effects of incubation temperature on the time required for incubation.

TEMPERATURE °C	INCUBATION PERIOD (days)
1	410
5	82
10	41

a) Why is fertilisation performed "in vitro" in trout hatcheries?
b) Use the data above to describe the relationship between the temperature and the incubation period for trout eggs.
c) Use your answer to b) to estimate the incubation period of eggs at the following temperatures:
   (i) 2°C
   (ii) 7°C
   (iii) 12°C
   (iv) 15°C
d) In the wild, eggs may be fertilised any time from late winter to early spring. The newly hatched fry are very susceptible to predation. Explain how the relationship described in parts b) and c) will aid the survival of the natural trout.

7 Hypophysation is a technique used to advance spawning in cultured fish.
a) Describe the mechanisms which normally control spawning.
b) Describe how hypophysation is carried out.
c) What is the advantage to the producer of being able to advance spawning in cultured fish?

8 A study was made of one particular species of fish in a freshwater lake. The fish were netted, measured and then returned to the water. The numbers in each length group were recorded as in the following histogram:

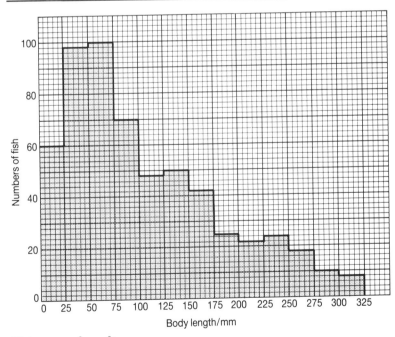

Histogram of results

a) Discuss the above data, suggesting possible explanations for, and limitations of, the given information when drawing conclusions.

b) Suggest two additional pieces of information that the investigators might usefully have recorded when carrying out their survey of the fish in order to make the data more scientifically useful.

9 A comparison was made of the size of trout of different ages in two lakes in North Wales. The average length of trout from a mountain lake, Llyn Mynydd, was compared with trout from Llyn Alaw, a reservoir in a lowland agricultural area. The results are shown in the table below:

AGE (years)	LENGTH OF TROUT (cm)	
	mountain lake	reservoir
1	5	12
2	10	25
3	20	38
4	21	42

a) (i) How could the age of the fish be determined?
    (ii) Give one reason why length was used as a measure of growth rather than any other parameter.

b) (i) Comment on the similarities in the growth patterns shown by the two populations of trout.
    (ii) What is the percentage difference in growth rates as shown by the average length of the two populations during their third year? Show your workings.

c) The difference in the growth rates of the two populations of trout has been attributed to the difference in nutrient availability in the mountain lake and the reservoir. Suggest how this difference could have arisen and why it would have an effect on the average length of the trout?

d) Suggest one other factor which could differ in the two lakes and explain how this might affect the growth rate of the trout.

ULEAC 1992

BIBLIOGRAPHY

Bardach, J.E. (1972) *Aquaculture.* Wiley Interscience.

Greenberg, D.B. (1960) *Trout Farming.* Chilton Co.

Love, (1970, 1980) *The Chemical Biology of Fish Volumes I and II.* Academic Press.

Reay, P.J. (1979) *Aquaculture – Studies in Biology No106.* Edward Arnold.

9 INVERTEBRATES – PESTS AND PRODUCERS

9.1 PESTS OR PRODUCERS?

The vast majority of animals known are invertebrates. They have a significant effect on our lives and many are of economic importance as either pests or as producers. Invertebrate pests can cause significant damage to both plants and animals:

- aphids and nematodes damage crops and reduce yields,
- parasitic organisms, such as tapeworms live inside the body of animals, taking nutrients from the animal itself and causing disease,
- mites and cockroaches feed on stored grain and flour causing damage and loss of yield,
- fleas, houseflies, locusts and aphids carry and transmit diseases to both animals and plants.

However, many invertebrates have useful roles:

- certain invertebrates, such as blowfly maggots, are responsible for the decomposition and decay of dead organic matter and the removal of this from the earth's surface,

- soil-dwelling invertebrates, for example earthworms, help produce and maintain a good soil structure which allows roots to grow successfully,
- invertebrates are responsible for bringing about pollination in some species of plant. Without them reproduction would be impaired,
- some specialised invertebrates are responsible for the production of highly marketable products, such as the production of both honey and wax by bees.

This chapter examines the impact on our lives of two species of insects — the honey-bee and the aphid.

9.2 THE HONEY-BEE

9.2.1 The honey-bee colony

Honey-bees are social insects. They live in large colonies which are highly organised and each honey-bee is adapted to perform a particular function within the colony. The colony is made up of three types of bee:

- the queen, a female bee whose function is to produce eggs for the production of new individuals,
- the workers, who are also females, act as "wet nurses" helping to rear the young, build the honeycomb in which the young are reared and collect food for the queen,
- the drones, who are male, mate with the queen.

A queen, worker and drone honey-bee

Each member of the colony, therefore, is dependent on others within the colony. By the middle of the summer the colony may consist of up to 50 000 bees — one queen, several hundred drones and thousands of workers.

9.2.2 Building a nest

In the wild, bees will build nests in a variety of places, including under window ledges, in hollow trees and under branches. In commercial bee production, the colonies are provided with a hive. This consists of a series of rectangular frames which sit inside a hive (see figure 89).

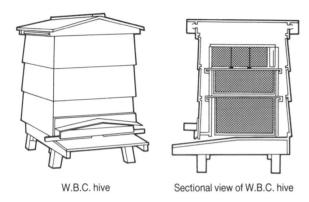

| W.B.C. hive | Sectional view of W.B.C. hive |

Figure 89 WBC (W.B. Carr) hive

The bee's nest is made up of a number of combs, which consist of hexagonal cells. The combs are constructed by the bees using beeswax which is produced by their bodies. They usually start at the top of the frame and build downwards. The hexagonal cells will house the developing bees and store honey and pollen. There are two types of cell: drone cells which are the larger of the two, and worker cells. Honey is stored in both types of cell, however pollen is stored in worker cells only. Cells for queen production are built as they are required.

Honeycomb

The queen will lay her eggs, one per cell. The egg is attached to the cell at one end. When hatched the larva is fed by worker bees for several days. At the end of the larval stage, the cell is capped. The pupa will continue to develop inside the capped cell until it emerges as an adult. Cells containing developing bees are called "brood cells" and appear dark in colour.

Brood cells are dark in colour compared with cells containing honey

9.2.3 The queen

Each colony of bees must contain one queen bee. The queen is much larger than the other bees within the colony, and is wasp-like in shape. She is responsible for the production of eggs and at the height of the season can produce between 1500–2000 a day. These eggs will develop into either drones or workers (see later).

The queen bee (centre) compared with workers

New queen bees are reared only under exceptional circumstances:

- if the old queen dies,
- if the old queen's laying ability begins to fail due to age or injury (**supersedure**),
- if part of the colony swarms, leaving the nest, thus dividing the colony in two (see later).

New queens develop in large cells which are specially constructed for the rearing of these bees. The cell has a thick wall, is about 3 cm long and hangs downwards. The number of cells produced will depend on the reason for their production. If the colony is preparing to swarm, many cells may be produced, however if the colony merely wishes to replace the existing queen only one or two cells may be built.

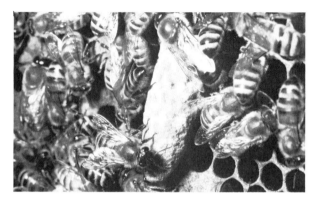

Queen cell

Fertilised eggs are deposited within the queen cells. The developing larvae are fed on **royal jelly**. This has a cream-like consistency and has a high nitrogen content. Royal jelly is produced by the worker bees and is fed to all developing larvae. However, drone and worker larvae are only fed it for about three days after which levels are reduced. The queen larvae are continuously fed royal jelly and so develop fully.

The virgin queen emerges from the cell by cutting a hole through the sealed end. In cases of supersedure, the old and new queen often work side by side, but if more than one virgin emerges at once they will battle until one is destroyed. Queen cells are often produced in succession so that the larvae are at different stages of development. If queen cells contain virgins which are nearly ready to emerge, the new queen will attack and destroy the occupants. However, if the cells contain larvae in the early stages of development, they will be left undisturbed. The successive production of queen cells allows the colony to split and swarm more than once if necessary.

After about a week, the virgin queen will take her "nuptial flight" during which she will mate with several drones. On returning to the hive, she will become the new laying queen.

9.2.4 The worker

The workers are the smallest but most numerous members of the colony. During the breeding season, numbers may exceed 30 000. Worker bees are female, but their reproductive organs are undeveloped and so they are usually unable to lay eggs. However, under certain conditions, eggs are produced by some workers. These develop **parthenogenetically** (without the egg being fertilised) into either queens or drones (see later). This process is important if the colony becomes queenless.

Worker bees

The workers have a fairly short life span depending upon the season and how hard they work. At the height of the season, worker bees will live for about 40 days. Workers are responsible for collecting nectar and pollen and storing it within the honeycomb. The honeycomb is made from wax which is secreted from glands on the abdomen of worker bees. Workers are also responsible for rearing and feeding the developing larvae with the royal jelly which they produce, and later with pollen and honey.

9.2.5 The drone

Drones are male bees. They are larger and stouter than the queens and workers, although the queen is longer. They have no sting, do not secrete wax or have suitable mouthparts for collecting nectar or pollen. This means that they are unable to carry out most of the functions performed by the workers. The main purpose of the drone is to impregnate the queen.

Drone bee

Drones are produced from unfertilised eggs by parthenogenesis. An unmated queen or, on occasion a worker bee, may produce eggs which are unfertilised by sperm. These are deposited into larger cells within the honeycomb and develop into male drones.

Drones are not present within the colony throughout the year, they appear in April or May. When the virgin queen emerges from the nest on her nuptial flight, the drones will follow and several will mate with her in mid air. Once mating has occured the drones will die. The remaining drones will be driven from the nest by the workers when food becomes scarce.

9.2.6 Swarming

The mass movement of bees is called **swarming**. Swarming is an important part of the bee behaviour as it prevents overcrowding within the nest or hive.

Bees collect nectar from flowers for food. During the spring, there is plenty of nectar available in blossom. This allows the bee population to increase. By the middle of May or June, the blossom season is over. The bee colony has grown to such an extent that more space is required. The colony will send out scouts which search for a place suitable to establish a new colony. When a suitable place is found the queen and a large proportion of the colony will fly off and start work on a new nest. The old nest is left queenless and unless it contains either virgin queens or queen cells containing larvae, has a limited lifespan.

Swarming is of obvious advantage to the bee colony, however is disadvantageous in commercial bee production as it may result in the producer losing his entire colony. Swarms can be recaptured once they have settled using a sealed container, and reintroduced to the hive. If a second swarm is to be prevented it is essential that the colony only contains one queen and that all others, including those developing within the queen cells, are removed.

Swarming

9.2.7 Honey production

Bees produce honey from the nectar they collect. Honey is the sole food source for adult bees. It is stored within the nest for use during the winter. Nectar is a sweet liquid composed of about 40% sugar (sucrose) and 60% water. It is produced by the nectaries of some flowering plants in order to attract bees and so aid pollination (see later). Worker bees collect both nectar and pollen from plants. Some of the pollen is stored within the nest and used as food for the developing larvae. If a plant contains large amounts of nectar, the bee may perform a dance around the flower. The bee will shake her abdomen from side to side and fly backwards and forwards in front of the flower. This dance attracts other bees to the bountiful plant so that nectar can be more completely extracted. Nectar is collected from the flower by foraging bees and is transferred from these bees to a house bee which returns to the nest carrying the nectar.

The house bee will convert the nectar into honey. This process involves two stages:

(i) The sugar in nectar is in the form of sucrose. This is converted by inversion into two simple sugars, glucose and fructose.

(ii) Water is removed from the nectar by evaporation. This occurs both during the manipulation of the nectar by the bee inside the hive, and during storage. The water content will fall from 60% to 20% over three days.

The honey is then placed into a cell in the honeycomb and sealed with a cap of wax. The honey will ripen within the cell. It takes the combined efforts of about 600 bees to produce 1 kg of honey. For each kilogram of honey removed from the hive, the bees will have produced and consumed a further 2 kilograms.

Honeycomb containing honey

Honey provides a rapid source of energy as it contains the simple sugars dextrose and fructose. These can be directly absorbed into the blood stream from the small intestine, without further digestion. Honey also has a higher mineral content than sugar.

> **Now try Investigation 16 Identifying the Sugars in Honey and Investigation 17 Measuring the Sweetness of Honey in _Animal Science in Action Investigations_.**

9.2.8 Beeswax production

Beeswax is secreted from four pairs of wax glands on the underside of the worker bee. Wax is used to construct the honeycomb cells in which honey and pollen are stored and eggs are laid. Clusters of worker bees group together causing the temperature to rise. Pentagonal-shaped wax scales form during the next 24 hours. These are used to construct the honeycomb cells. They are stuck together using saliva. The hexagonal cells are constructed so that they are closely packed together. The cell is hollowed out by the bee which removes wax from the sides of the cell until it can reach the bottom. Beeswax has a wide variety of uses. Some are listed in table 9.1 below.

Table 9.1   The uses of beeswax

USE	EXAMPLES
Adhesives	Adhesives which stick glass to glass, metal to glass, adhesives for wig-making
Candles	Church candles, decorative candles
Cosmetics	Grease paint, cleansing creams, eyebrow pencils, lipstick
Crayons	Wax crayons, drawing pastels
Food	Honeycomb, sweets, decorations
Oils	Axle grease, lubricants
Paints and varnishes	Paint, paint and varnish removers
Paper	Coating of washable wallpaper, carbon paper
Pharmaceutical	Hair restorer, shaving cream
Cleaning materials	Floor polish, shoe cream, car wax
Textiles	Waterproofing of canvas, cellulose etc.

9.2.9 Honey-bees as pollinators

The honey-bee is of great economic importance as a pollinator of crop plants. Many flowering plants are pollinated by bees in nature, however the bee is invaluable in commercial apple production. The apple is self-sterile which means that it cannot pollinate itself. It requires crossing with pollen from other species. Bees will pick up pollen from apple trees when collecting nectar. This is then transferred to the flowers of another tree from a different species as the bee collects nectar. This causes cross-pollination and enables the tree to produce fruit. Honey-bees are also used as pollinators inside greenhouses to pollinate crops such as cucumbers. Table 9.2 below lists some important crops which are pollinated by bees.

Cross pollination by bees

Table 9.2   Important crops which rely on pollination by bees

Fruit crops	Apple, apricot, blackberry, cherry, grape, peach, pear, plum, raspberry, strawberry
Seed crops	Broccoli, brussels sprouts, cabbage, carrot, cauliflower, celery, cotton, onion, parsnip, pepper

9.3   THE APHID

Aphids are small, soft-bodied insects which feed on sap extracted from plants. There are over 4 000 species of aphid, mainly living in temperate regions. In a hectare of land where plants are growing, there may be up to $5\,000 \times 10^6$ aphids feeding on leaves and shoots and another 650×10^6 on the roots. Each species of aphid feeds on only one or two plant hosts, however, they are able to migrate great distances (1300 km) in search of food.

9.3.1 Aphid life cycle

The aphid has a very unusual life cycle. The adult can exist in two forms. At certain times of the year the adults are able to reproduce sexually and at other times they reproduce by **parthenogenesis**. Aphids also feed on two plant hosts depending on the season and stage of their life cycle. The life cycle may vary slightly from species to species, but the majority of aphid species have life cycles similar to that of the black bean aphid (*Aphis fabae*).

The black bean aphid's eggs overwinter on the spindle tree (the primary host). In late May they begin to hatch. These eggs are produced as a result of sexual reproduction. The resulting aphids may be wingless at first, however, as numbers increase and food becomes scarce, winged individuals develop. These individuals are able to migrate to the secondary host – in this case this is often the black bean but it may also be the poppy, sugar beet or dock. Once the aphids have found a suitable host, they will start to produce young by parthogenesis. These offspring will initially be wingless and all females. The females continue to produce live young. This method of reproduction is very efficient because it avoids the time required for egg production and embryo development. The live young are able to start feeding immediately. Figure 90 shows the potential numbers of descendants from a single female using (A) parthenogenesis and (B) sexual reproduction. When food is scarce, winged females develop and migrate to find new secondary host plants.

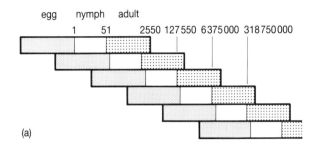

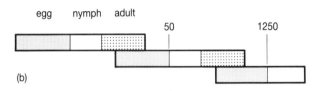

Figure 90 A comparison of the number of descendants produced by (a) parthenogenesis and (b) sexual reproduction

As autumn approaches, male and female winged aphids are produced. These are able to reproduce sexually and migrate back to the primary host – the spindle tree. These aphids will mate on arrival and the female will lay her eggs on the stems and winter buds of the tree. Each female produces 4–6 eggs which are shiny black in colour. The parents will die shortly after the eggs are laid, leaving them to overwinter and hatch in the spring.

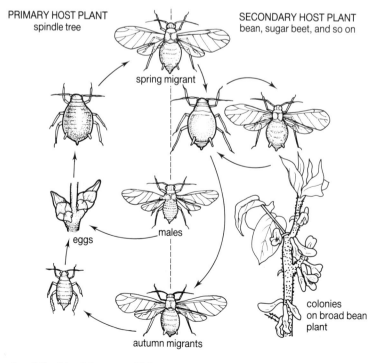

Figure 91 The life cycle of the black bean aphid

Now try Investigation 18 Measuring the
Reproductive Rate of the Aphid and
Investigation 21 Selecting a Host in *Animal
Science in Action Investigations.*

9.3.2 Feeding in the aphid

Aphids feed through their specially designed
mouthparts. The mouthparts consist of two
mandibles and two maxillae which are extended
to form **stylets.** The stylets have many functions.
The mandibular stylets help to pierce the leaf
surface. Saliva is pumped into the leaf through
one of the maxillae stylets. This contains
pectinase and other enzymes which help to soften
the leaf tissue. Once the phloem has been
penetrated, the other maxillary stylet sucks up the
plant sap.

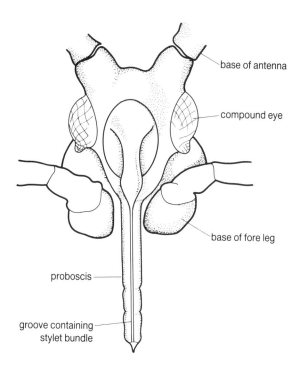

Figure 92 Aphid mouthparts

Plant sap contains large quantities of sucrose.
Most of this passes through the aphid where it is
converted into trisaccharide sugars and excreted
as **honeydew.** Honeydew is used by free-living
nitrogen-fixing bacteria in the soil, making more
nitrogen available for the plant.

**Now try Investigation 19 Measuring the
Feeding Rate in Aphids and Investigation 20
Examining Honeydew in *Animal Science in
Action Investigations.***

9.3.3 Aphids as crop pests

Individually, aphids cause little damage to their
hosts, however, each colony may contain
thousands of insects and this can cause problems.
Aphids affect the yields and quality of several
crops. For example *Aphis fabae* reduces the
numbers of seeds produced by the bean plant by
80% and the average mass of the beans by 45%.

Aphids act as parasites, feeding on the plant's
nutrients. This often stunts the growth of the plant
or even causes death. Some aphids have toxic
substances in their saliva which cause the leaves
to curl or galls to be produced. The most harmful
effect aphids have on plants, however, is in
transmitting viral infections.

Aphids transmit viral infections in two ways:

- virus particles are carried on the tips of the
  stylet and injected directly into the phloem
  during feeding,
- the virus infects the aphid's blood, circulates
  around the body and is injected into the plant
  in the saliva during feeding.

Stylet-borne viruses, for example potato virus Y,
are transmitted easily as the aphid may not even
need to feed on the plant itself. However, these
types of viruses are lost when the aphid next
moults.

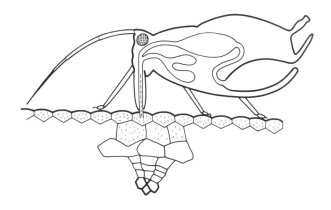

Figure 93 Stylet-borne viruses

Circulatory viruses are more difficult to transmit. They multiply inside the aphid for anything up to several weeks. They are then transmitted to all plants which the aphid feeds on throughout its life.

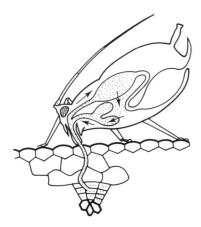

Figure 94 Circulatory transmission of viruses

9.3.4 Aphid control

The rapid reproduction rate of the aphid means that a population size can soon reach serious proportions. Under natural conditions, population size is controlled by factors such as temperature, predators and the availability of food. Agricultural systems, however, cannot withstand large aphid populations and so the numbers must be controlled. This can be done by the use of chemical pesticides or natural enemies.

Agrochemicals are extremely effective and they have an excellent safety record, (B.A.A. 1991), but there is growing public concern about the use of chemicals in food production and the financial cost is very high. As a result of these pressures there is increasing interest in alternative control measures as we will see later. Many of the chemical pesticides used today are selective in their action and so are effective if reducing the pest population but do limited damage to other species. Aphids are most effectively controlled by the use of **systemic pesticides**. Systemic pesticides are absorbed through the roots and leaves of plants and translocated around the plant. The plant's sap becomes toxic to the pest and so when the crop is attacked by the pest the poison is taken directly into its system. Systemic pesticides tend to have longer term effects as they remain in the plant. They have a protectant effect on the crop as plants can be treated with pesticides to control pests which are likely to cause infestations before they occur.

In more recent years there has been great interest in using "natural enemies" to control pests. This method is particularly successful in the control of insects and relies on the fact that most common pests are prey for other insects.

Aphids are common pests of gardens, greenhouses and crop plants. They are, however, prey for hoverfly larvae which possess piercing mouthparts and spear the aphid, sucking out its body fluids.

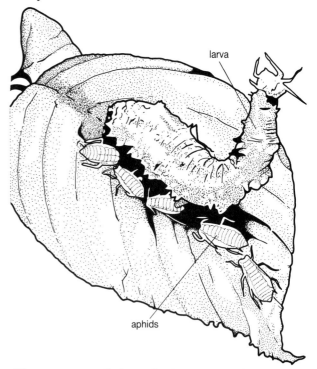

Figure 95 Hoverfly larva feeding on aphids

Adult hoverfly

In the 1980s, the Department of Entomology and Insect Pathology produced the following data. Adult hoverflies were introduced into controlled environments containing aphid-infested cucumber plants. The adults were allowed to lay their eggs and then were removed. The larvae developed from the eggs and started to feed on the aphids. The results obtained are shown in figure 96.

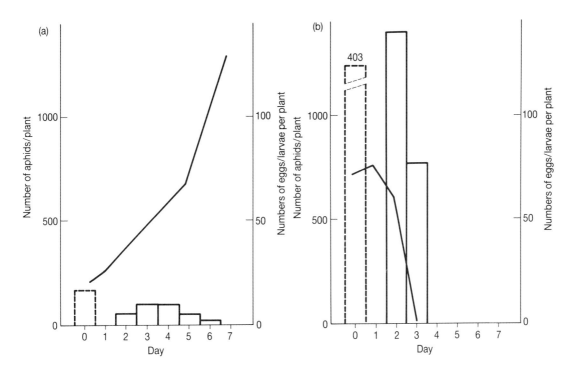

Figure 96 Graphs showing the effects of adding hoverfly eggs to a population of aphids

Graph (a) shows the effects of a small number of hoverfly larvae on the aphids. If the larva population size is too small then the aphids are able to survive this method of control. However, if the hoverfly larvae are introduced in sufficient numbers then they have a dramatic effect on the population of aphids.

This method of pest control would seem to have many advantages over the use of chemicals. However, it is an unsuitable method of control for many pests as it relies on the fact that a natural enemy exists for the pest. It is also impossible to use this method of control in the field environment.

Another method of biological control of pests involves the release of males which have been sterilised by irradiation treatment. If sterile males are released into the environment, the females which mate with them lay eggs which have not been fertilised and so do not develop. Eventually the insect population will be destroyed as adult insects are not replaced.

Integrated pest management (IPM) is a pest control technique which was developed in the 1960s. The main aim of this technique is to control the pest population so that it is not totally destroyed but so that their numbers are below the level which causes uneconomic damage. Integrated pest management usually involves the use of non-chemical control methods e.g. natural enemies (biological control) or physical control, with the application of chemical pesticides only when the pest population exceeds a certain level.

When introducing this technique a grower must firstly be aware of the types of pest which are likely to attack his crop and select a range of non-chemical methods to control them. This may involve the use of natural enemies. If these are to be encouraged it may be necessary to provide them with a suitable habitat or an extra source of food. The pest population must be monitored carefully so that the population size does not reach levels which will cause crop loss. This is time consuming and can be difficult to assess, however IPM results in a decrease in pesticide use. This reduces costs and has environmental advantages.

QUESTIONS

1 The honey-bee colony is highly organised.
  a) Briefly explain the roles of the following members of the colony:
       (i) The queen,
       (ii) The drones,
       (iii) The workers.
  b) The honey-bee nest contains only one queen. Under what conditions are new queens produced?
  c) Why is swarming important to the honey-bee colony and a disadvantage to the farmer? What measures can be undertaken to prevent swarming?

2 The graph below shows the incidence of the beet yellows virus and the population size of two species of aphid.

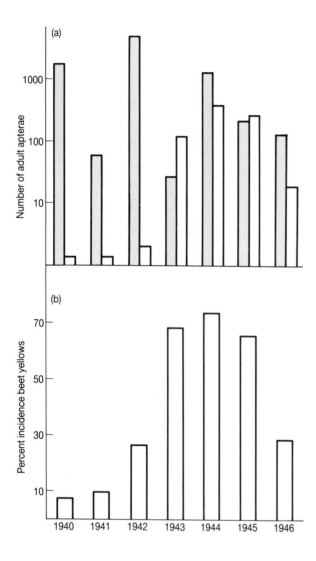

The transmission of beet yellows by aphids

Beet yellows virus is a disease of sugar beet which can be transmitted by both the peach-potato aphid and the black bean aphid. The black bean aphid is found in large numbers but colonies tend to aggregate on only a few plants at a time. The peach-potato aphid is far more restless and although it is present in smaller numbers, colonies tend to move quickly throughout a crop.

a) Aphids can transmit viruses in two ways. Using diagrams, explain how
  (i) stylet-borne and
  (ii) circulatory viruses are transmitted.
b) Which of the two species appears to be most successful at transmitting beet yellows virus? Explain your answer.
c) Why are systemic pesticides particularly effective at controlling aphids?

3 Aphox and Metasystox are two commonly used insecticides which control aphids in cereal crops.

Table	A comparison of Aphox and Metasystox	
INSECTICIDE	COST	SELECTIVITY
Aphox	£15/ha	Aphids only
Metasystox	£7.50/ha	Aphids, ladybird, capsid wasps and others

What are the advantages and disadvantages of using each of these insecticides to control aphids in a cereal crop?

4 The diagram below shows the relationship between the movement of the bird cherry oat aphid from the bird cherry to the oat in early summer and back again in the autumn, and the percentage of soluble nitrogen present in the leaves of the bird cherry tree.

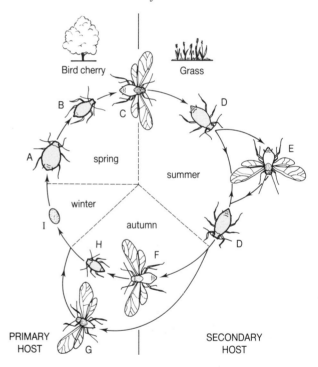

The migration of the bird cherry aphid

a) Use the information in the graph to construct a diagram showing the life cycle of this aphid. The diagram should indicate the type of reproduction which occurs at each stage.
b) What appears to cause the migration of the aphids in this example from their primary to their secondary hosts and back again?
c) Aphids feed by sucking sap from their host. Honeydew is produced as a waste product. What is the advantage of this secretion to the host plant?

5 The diagrams below show the three types of honey-bee, labelled A, B and C.

(a)

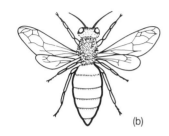

(b)

(c)

Honey-bees

   a)   Identify A, B and C.
   b)   State three ways in which bees are economically important.

ULEAC 1992

6 What are the main advantages and disadvantages of integrated pest management?

7 a)   Make a labelled diagram to illustrate the life cycle of an aphid.
   b)   State two ways in which aphids are of economic importance.

ULEAC 1992

BIBLIOGRAPHY

Blackman, R. (1974) *Aphids.* Ginn and Company Ltd.

Dixon, A.F.G. (1985) *Aphid Ecology.* Blackie and Son.

Dixon, A.F.G. (1978) *The Biology of the Aphid* (Studies in Biology no 44). Edward Arnold.

Free, *The Social Organisation of the Honey-bee* (Studies in Biology no 81). Edward Arnold.

Mace, H (1952) *The Bee Keepers Handbook.* Ward, Lock and Co.

Wigglesworth, V.B. (1984) *Insect Physiology.* Chapman and Hall.

GLOSSARY

acrosome: enzyme-coated tip of the sperm head.

actin: protein found in muscle.

actinomyosin: protein complex formed when actin and myosin combine during muscle contraction.

albumen: the white of the egg, composed of mainly water and protein (albumin).

allantois: extra-embryonic membrane found in eggs that functions as an extra-embryonic lung and kidney.

amnion: extra-embryonic membrane found in the chick eggs that forms a fluid-filled cavity which helps protect against mechanical damage and temperature and humidity changes.

amniotic fluid: fluid which fills the cavity surrounding the growing foetus.

aquaculture: rearing fish for food and sport on fish-farms.

artificial insemination: introduction of sperm into the vagina of the female by artificial means.

battery egg production: the intensive production of eggs by hens that are housed in cages.

biomass: the dry mass of all the living organisms in a given population or given area.

blastocyst: layer of cells produced soon after an egg is fertilised by the sperm.

blastodisc: position at which the sperm enters the egg during fertilisation.

blastula: first stage of embryonic development.

breed: different variety of the same species of animal.

broiler: chicken reared for meat production.

butterfat: fat content of milk.

candling: technique used to determine the quality of egg shell.

cartilaginous fish: fish with a skeleton made up of cartilage, for example dogfish and sharks.

chalazae: twisted cords of albumen which hold the egg yolk in a central position.

chimeras: an organism made up of two genetically distinct tissues.

chorion: extra-embryonic membrane in a chick egg.

cloaca: a digestive and urino-genital opening found in most vertebrates, except mammals.

clone: individual which is genetically identical to another and is derived from the same zygote.

co-dominance: results when neither form within a pair of genes is dominant to the other. This leads to a combining of characteristics to produce an intermediate form.

colostrum: the first milk produced by mammals just after the birth of young. This milk is rich in nutrients and contains antibodies to help the young animal fight infection.

demersal fish: fish living at the bottom of the sea.

dominance: used in reference to the form of the gene pair that is expressed if present in an individual.

drone: male bee.

ejaculation: release of semen from the penis.

electroejaculator: machine used to stimulate semen release during artificial insemination.

embryo transfer: technique during which ova are removed from one animal, fertilised externally and then placed into the uterus of another animal where they will grow and develop.

epistasis: used in reference to inheritance to describe the situation when one gene controls the expression of another.

fecundity: reproductive rate.

free-range eggs: eggs produced by chickens which are not housed in batteries. Living conditions must comply with Government and EC standards.

gastrulation: second stage of embryonic development.

gene transfer: the removal of desired genes from the DNA of particular cells and their introduction into the DNA of other cells.

genotype: the genetic make up of an individual.

heifer: female cow which is yet to produce young.

heritability: the extent to which a given genetic characteristic is passed onto offspring.

heterozygous: containing two forms of a particular gene.

homozygous: containing the same forms of a particular gene.

hybrid: an individual produced as a result of the crossing of two different strains.

hypophysation: the artificial stimulation of ovulation by hormone injection.

inbreeding: mating of individuals of the same species and of the same breed.

invertebrate: animal without a backbone.

lactation: milk production.

letdown-reflex: reflex by which milk begins to flow.

meiosis: cell division associated with gamete formation.

mitosis: cell division associated with growth.

myosin: protein found in muscle.

nectar: substance collected from flowers by bees and used to produce honey.

oestrous cycle: hormonally controlled reproductive cycle of a female animal.

oestrus: period of "heat" during the oestrous cycle in which mating can occur.

outbreeding: mating of individuals within the same species but of different breeds.

oogenesis: egg formation.

oxytocin: hormone involved in the let-down of milk.

parthenogenesis: the asexual production of young.

pasteurisation: the heat treatment of milk so that pathogenic bacteria are destroyed.

pathogen: disease-causing organisms.

pelagic fish: fish living near to, or at the surface of, the sea.

phenotype: the characteristics shown by an individual.

photoperiodism: responses to the length of day.

polygene: a collection of genes which have an additive effect in the inheritance of certain characteristics.

queen bee: the "head" of the honey-bee colony – the primary egg-laying bee.

Resazurin: dye used in the testing of milk quality.

Sertoli cell: cells which provide nutrients for the maturing sperm cells.

solids not fat: solid proportion of milk excluding the butterfat.

spermatogenesis: sperm production.

sterilisation: the heat treatment of milk to destroy bacteria.

stylet: specially adapted mouthpiece which allows sap to be sucked from a plant.

teleost fish: bony fish.

transgenic animal: animal which has genes from another source incorporated into its DNA.

vitelline membrane: membrane which surrounds the yolk of an egg.

worker: female bee, usually does not lay eggs but is responsible for feeding larvae, producing honey and beeswax.

yolk: the yellow of the egg – provides food for the developing chick embryo.

INDEX